EDGES OF BOUNTY

EDGES  OF BOUNTY

Adventures in the Edible Valley

Edibilism *n*: The common human desire to personally participate in the production of some or all of one's food or drink. Intentional practitioners of edibilism, including gardeners, farmers, butchers, canners, wine makers, fishers, gatherers, and hunters, are edibilists.

Words by William Emery
Photographs by Scott Squire

Foreword by NPR's Kitchen Sisters,
Davia Nelson and Nikki Silva

Heyday Books, Berkeley, California
BayTree Books

This book was made possible in part by a generous grant from the BayTree Fund.

"Turnip Crop" by Rolf Jacobsen, translated by Robert Greenwald, from *North in the World: Selected Poems of Rolf Jacobsen*, The University of Chicago Press, 2002. English translation © 1985, 1997, 2002 by Robert Greenwald. Reproduced by permission of the publisher.

Library of Congress Cataloging-in-Publication Data

Justice, William E.
  Edges of bounty : adventures in the edible valley / words by William Emery ; photographs by Scott Squire.
    p. cm.
  "Edibilism: The common human desire to personally participate in the production of some or all of one's food or drink. Intentional practitioners of edibilism, including gardeners, farmers, butchers, canners, wine makers, fishers, gatherers, and hunters, are edibilists."
  "BayTree Books."
  ISBN 978-1-59714-108-6 (pbk. : alk. paper)
 1. Farms, Small--California--Anecdotes. 2. Family farms--California--Anecdotes. I. Squire, Scott. II. Title.
  HD1476.U62C354 2008
  338.1097945--dc22
                          2008008689

Cover Photograph: Tomato truck, Yolo County
Book Design: Lorraine Rath
Printed by Global Interprint in Singapore

Orders, inquiries, and correspondence should be addressed to:
        Heyday Books
        P. O. Box 9145, Berkeley, CA 94709
        (510) 549-3564, Fax (510) 549-1889
        www.heydaybooks.com

10 9 8 7 6 5 4 3 2 1

Contents

List of Plates   vii

Foreword by The Kitchen Sisters, Davia Nelson and Nikki Silva   ix

Introduction   xi

Section One: Farmers

The Bearer of Strange Melons   3

Wherever People Are Hungry   9

Famous Everywhere but Brentwood   19

I Am a Jam Maker   25

Roadside Stands: A Pastiche   33

Section Two: Milk, Meat, and Honey

The Lopez Lamb Company   61

Save the Star Thistle!   67

Like Sharks on Bloody Meat   75

Bubbas   83

Red, White, and Roan   91

Section Three: Fish

River Rats Want Sugar in Their Coffee   99

A Day on the Delta   107

That's Why They Call It Fishing   111

Section Four: The Kitchens

Tucos Wine Bar and Café   123

Sopash   129

The Wool Growers Rest   137

HuntFishCook   143

Aisu Pops   149

Section Five: Found on the Road

Volta, Unincorporated   157

The Egret Palace   161

Let My People Go   165

Section Six: Killing, the Beginning, and the End

Chico Lockers and Sausage   173

Junior and Ladies Pheasant Hunt   179

Everyone Should Know Everything   189

List of Plates

1.1   Topsoil, Contra Costa County

1.2   Shovel at rest, Yolo County

1.3   Mike Madison shares a melon, Yolo County

1.4   The Madison kitchen, Yolo County

1.5   Shelling walnuts, Yolo County

1.6   Knife and nopales, Yolo County

1.7   Ursine sleepers, Yolo County

1.8   Cardoon, Contra Costa County

1.9   Packing rosemary, Contra Costa County

1.10 The nursery, Contra Costa County

1.11 Burst watermelon, Tulare County

1.12 Flash-frozen fruit, Tulare County

1.13 Paul Buxman at home, Tulare County

1.14 Groppetti's Deli, San Joaquin County

1.15 Old chicken coop, Butte County

1.16 Tavern Pale, East Nicholas, Sutter County

1.17 Yosemite-bound, Stanislaus County

1.18 Fallen strawberry, Fresno County

1.19 Taco truck, Knights Landing, Yolo County

1.20 Stone fruit, Butte County

1.21 Mass grave, Yolo County

1.22 Tomato truck, Yolo County

1.23 Dead kiwi vineyard, Tulare County

1.24 Three onions, Yolo County

1.25 Strawberries for sale, Fresno County

1.26 Psalm book, Fresno County

1.27 Overripe fruit, Butte County

1.28 Last year's okra, Tulare County

1.29 Award-winning jam, Sacramento County

2.1   Adopted turkey, Tuolumne County

2.2   Trophy room, Sacramento County

2.3   Team Lopez, Sacramento County

2.4   Herding, Sacramento County

2.5   Honey for sale, Contra Costa County

2.6   Milking the goat, Contra Costa County

2.7   Wax-handled hammer, Contra Costa County

2.8   Hives, Sutter County

2.9   Honey gut, Sutter County

2.10 Safflower field, Sutter County

2.11 Twins, San Joaquin County

2.12 Women at auction, San Joaquin County

2.13 Welcome, San Joaquin County

2.14 Lonely calf, San Joaquin County

2.15 Red, white, and roan, Solano County

2.16 Buckets, Solano County

2.17 Spilled milk, Solano County

3.1   Bass fisherman, San Joaquin County

3.2   Permanent shanty, San Joaquin County

3.3   Fishing hatch, San Joaquin County

3.4   Hezbollah is here, San Joaquin County

3.5   Hyde Street grip, San Joaquin County

3.6   Dike fishing, San Joaquin County

3.7   The catch, San Joaquin County

3.8   Samurai, San Joaquin County

3.9   Abandoned boat, Sacramento County

3.10 Grand old flag, Sacramento County

3.11 Morning fisherman, San Joaquin County

3.12 Hidden bass, San Joaquin County

4.1   Tucos, Davis, Yolo County

4.2   On the vine, Colusa County

4.3   Nevermore eggs, Colusa County

4.4   Tree art, Colusa County

4.5   Sopas warrior, Buhach, Merced County

4.6   Open kitchen, Buhach, Merced County

4.7   Calm before the storm, Buhach, Merced County

4.8   Queens, Buhach, Merced County

4.9   Bread and mint, Buhach, Merced County

4.10 Wool Growers Rest, Los Banos, Merced County

4.11 Obeisance, Los Banos, Merced County

4.12 Aftermath, Los Banos, Merced County

4.13 Overcooked duck breast, Sacramento County

4.14 Herb garden, Sacramento County

4.15 Pop city, Davis, Yolo County

4.16 Cantaloupe, Davis, Yolo County

4.17 Jaymes' Best Pops, Davis, Yolo County

4.18 Tricycle, Davis, Yolo County

5.1 Downtown, Volta, Merced County

5.2 Lambs on the lawn, Volta, Merced County

5.3 Rosie, Volta, Merced County

5.4 "No Fishing," San Joaquin County

5.5 Scarecrow, San Joaquin County

5.6 Abandoned bait shop, Knights Landing, Yolo County

5.7 "Let my people go," Sacramento County

6.1 Deer legs, Galt, Sacramento County

6.2 Meat locker, Chico, Butte County

6.3 Small buck, Chico, Butte County

6.4 Cured ham, Galt, Sacramento County

6.5 Shot pheasant, Sacramento County

6.6 Bird dogs, Sacramento County

6.7 Young hunter with dog, Sacramento County

6.8 Pheasant hearts, Sacramento County

6.9 Skinned pheasant, Sacramento County

6.10 The Silva kitchen, Sacramento County

6.11 Leeched acorn meal, Madera County

6.12 Fallow deer skin, Madera County

6.13 Ron and Elena Littlefawn Goode, Madera County

6.14 Pan-fried catfish, San Joaquin County

Foreword

The Kitchen Sisters, Davia Nelson and Nikki Silva

THE KITCHEN SISTERS SEARCH for our hidden kitchen cultures, for the people who fly below the radar who glue their communities together through food. So do William Emery and Scott Squire. Only they have gone on a quest for what is in plain sight, or used to be anyway, a roadside fruit stand, a field well tended. They make house calls, two observers, on the road, barreling through California's Central Valley on their version of Route 66. Imagine if Buzz and Todd were looking for roadside produce instead of chicks. If Dean Moriarty and Sal Paradise were checking out biodynamic farm practices instead of speed rapping about writing, jazz, and Denver. These guys are also on a quest, but in this case, it's the rapture of a peach in season, the grace of a melon patch and a long row of peppers, the wild philosophies of farmers, their visions, furrows, and passions.

The fellowship of food. That's what we call this interconnected world of people who feed and tend their communities with their crops and traditions. This collection looks deep into the heart of the peach, and the people who grow it. It taps into a longing and a passion for authentic taste and for knowledge of the lives of people who devote themselves to these fresh, baked, canned, and preserved endeavors. *Edges of Bounty* finds its characters in the outposts, on the margins, at the crossroads. Through the eyes of William and Scott, these people emerge as folk heroes, kitchen visionaries. These are individuals still tapped into land and place, rooted in their philosophies, their practices, their maniac desire to feed their families and the planet something healthy, gorgeous, and delicious. You read it in their words, you see it in their faces and the way they live and stand their ground. The land comes to life. The Central Valley of California is rendered again. Writer and photographer watching, asking, searching.

Look around you. Do you like what you see? On the land you see before you, what is going on? How about on the land where the food you eat is grown? Do you know the land where your food is grown? *Edges of Bounty* is perhaps a kind of road map. An invitation, a set of possibilities. It beckons. Find a friend, grab your camera and notebook. Hit the road and look for something. Start talking to strangers. Find what you're missing, needing, what gives life purpose. The quest, the journey that gives meaning. How do people resist? In the hands of the right person, growing a melon is an act of social resistance. Maintaining a fruit stand on a whizzing highway becomes the last stand against anonymous, corporate agribusiness—a way to survive, to connect, to keep the farm operating for one more season.

Edges of Bounty. That's what's left. They used to be called Pastures of Plenty. Now, these same farmers are forced to the margins—staking a claim with their weedpatch peppers and their cactus fences. United in fruit, in a love of land, in the deep need to grow, to tend and feed their communities. These are people on the Edges, on the knife's edge, Ramón's knife edge of survival, risking everything on the harvest and the land, on finding a market for their vision.

Who glues your community together through food, and what would you be eating if farming on a smaller scale ceased to be? If Ramón and Lucy, and Mike Madison, and Rick Knoll, and Paul Buxman, and Cherie with her trees and pies threw in the towel, just couldn't hack it anymore against the competition from the big guys. Are you growing anything, or cooking much these days? Are you wondering what the future of food is? It's simple and complicated. If it came down to it, could you provide for yourself and those you love? Are you using your land?

These are stories. Of guerrilla agriculture, and the fast-disappearing feral things, the wilder land and crops. The wilder and more untamed parts of ourselves that connect back to land and place. These are stories of orphans, gleaners, artisan farmers, dogpatch visionaries, agro-refugees, last standers. It's an homage to the scrappy and wild, to the humor and bodacious attitude that let people stick it out against all odds.

This collection is disorienting, disquieting, then tranquil and almost surreal. It is full of images that tug and point the way: the shimmer and glow of a long truck with a scattering of fallen fruit forgotten beside it on the roadside, a glowing red heart etched against a brown barn. Jars of honey and honeycombs resting on a porch. Three onions in the dirt.

The land is the last stand. Wendell Berry says, "Eating is an agricultural act." Agricultural acts are about eating. And about resistance, and commitment. And this collection knits those worlds together. It celebrates the ecstasy of season, spotlights the agony of chemicals and monoculture and agribusiness. It takes pleasure in the communion, and the passion and the fellowship. And turns its gaze to the Valley, where a farm stand is a last stand.

This book breathes. It is vivid, full of small revelations and deep meditation on land and environment. A series of portraits and reflections. Its heart, like ours, is the kitchen, where so many good stories begin and end.

Introduction

"All real living is meeting."
—Martin Buber

THERE IS A GREAT DEAL OF ANTAGONISM in the substratum of this book. In a sense, the Great Central Valley itself dictates a certain amount of strife. In fact it is two valleys, the Sacramento and the San Joaquin, and insofar as the word "valley" itself carries with it a sheen of pastoral simplicity, the Great Central Valley of California is anything but. It is one of the great engines of global capitalism and industry, and it has taken as its model not the orchards, villages, and black dirt fields of yore, but the factory floor. We traveled there, however, not for this story but its opposite. Geographically, it was like going to the North Pole to find the South. Politically, we were looking for the Resistance to mechanical reproduction: the human hand.

The length and breadth and depth of what we found caused a crisis in our language. By taking food, or the edible, as our subject and taking the measure of our subject with the human hand, we wove such disparate activities as beekeeping, artisan popsicling, ditch fishing, melon theft, pheasant hunting, and small farming into one fabric. Our publisher thought that there should be a word for what was instead an endless and endlessly varied list. I turned over a few agrarian rocks and stuck my hand in some luddite crannies and found nothing. Eventually, the word "edibilism" startled me out of a good night's sleep and adhered itself to our project.

An "ism," once wrought, will grow at terrible rates. What began as a term of convenience has become a layered and formidable critique of not only the global supermarket but of other reform movements in the food world. We have been made drunk by the supposed power of our own consumption, led to believe with some fanaticism that simply buying and eating the right food will lead us into a promised land of sustainability. By this logic, the more you consume, and the more you pay for the privilege of that consumption, the greater an agent of social change you are.

Two foodies meet on the street, both carrying Napa cabbage. The first one says proudly, "I paid sixteen dollars for this Napa cabbage." The second foodie says, "I've got you beat. I paid *twenty* dollars for the same cabbage up the street!"

And so, the victory of consumerism is so complete that it has even acquired our moral imagination. The radical thrust of edibilism is that it places the buying and selling of food farthest from its purpose, which is to return us to our heritage as makers and sharers. By making intimacy the primary ethic, edibilism has the ability to address the plight of our rural areas, or rural life itself. For until a thriving and autonomous rural life is not only possible but probable, there will be no just consumption of its produce.

For a moment, let's stand up and look around. The enormity of what lies before us cannot be avoided. The rural, wherefrom derives every necessary thing, has been enslaved, debased, and colonized by a parasitical urbanism. Worldwide, this assures that no community nor any culture will be able to escape empire, that no one will be free.

But we are no ideologues! This book is about people and the strange accident of their lives. To record this accident is the artist's chief duty—to follow orders, empty of intent, not knowing nor caring where the blade drives home. Scott's photography, which always preceded the text, set the standard. Neither landscaper nor portraitist, Scott creates, with his photographs, a language of meeting between the human and the natural. Rather than drugging us with the ecstasy of the human face or the grandeur of the horizon, his square film captures hands, tools, feet, and fabrics. In words and pictures, we present here images of rural life, where humanity and nature meet each other as equals and partners, a nourishing rather than annihilating mystery.

Turnip Crop

It's heavy, what we lift from the earth here
is heavy as bricks and marble pedestals,
remnants of the great sun-temple Huitzilopochtli.
Dig more, here are the goddess's thighs
dark with earth. Carry them off.
                                        —Rolf Jacobsen

PLATE 1.1

Let me hope the whole of that landscape we shall essay to travel in is visible and may be known as there all at once: let this be borne in mind, in order that, when we descend among its windings and blockades, into examination of slender particulars, this its wholeness and simultaneous living map may not be neglected, however lost the breadth of the country may be in the winding walk of each sentence.

—James Agee

Plate 1.2

ONE: Farmers

Plate 1.3

The Bearer of Strange Melons

WATCH OUT FOR TRACTORS.

Or, perhaps the glyph was more circumspect, meaning simply: TRACTORS.

As though we had forgotten they existed.

The sharp, yellow diamond standing sentinel along Putah Creek Road clearly meant to warn us of some peril as we paused before it, having just exited the concrete, rubber, metal, and plastic hell of Interstate 80, which roars across the entire United States from San Francisco, California, to Teaneck, New Jersey. But the sign's message was, for me, ambiguous. The black silhouette of the tractor was fat and friendly. The silhouette of a man sat atop the tractor, and atop the man sat the black outline of a well-worn straw hat. The smallness of the tractor and the nostalgic roundness of the figures evoked a landscape far different from what surrounded us. I already knew that the farm machines trundling down the small roads we had set out to explore would look nothing like the quaint tractor on the sign—most would not even be tractors. Of what, then, was this sign, so conspicuously placed within the first ten minutes of our journey, warning us?

It had all begun with signs, not the officious metal codes of conduct erected by the state, but with handpainted signs made of scrap wood meant not to warn but to entice.

FIRESTONE PEACHES   RIPE STRAWBERRIES

OKRA   MELONS   LOCAL HONEY

The Great Central Valley of California, which is the name given to the combined Sacramento and San Joaquin Valleys, running from Red Bluff down to Bakersfield in between the Sierra Nevada and the Coast Ranges, has been remade and celebrated in the image of "big" agriculture—called agribusiness and monoculture—and worked with sophisticated machinery and millions of migrant workers. Its reputation is one of drear and moral turpitude. But...

TOMATILLOS

FRESH EGGS

HOMEMADE PIE

Something else was going on in this place, we had realized, and we needed to find out what it was and who was responsible. The yellow sign on Putah Creek Road and the idea of farming, personal, warm, man-sized, that it heralded was perhaps really what we ought to be watching out for. Just then, we pulled off the road to let an enormous combine pass by, emitting enormously the dusty redolence of sunflower seeds.

MIKE MADISON FARMS flowers, melons, olives, persimmons, quince, and other foods for mouth and eye on twenty-two acres just forty minutes' fast driving from the San Francisco Bay. He has published two books: one, *Walking the Flatlands*, is a deep ecology of the land around the Lower Putah Creek, which forms the eastern edge of his property; the other, *Blithe Tomato*, is a series of short ruminations on farming, farmers, and farmers' markets. He was a Harvard-educated botanist and worked in South America before marrying, purchasing this land, and settling less than a mile from the farm on which he was raised. One of his sisters, Deborah Madison, is the author of *The Greens Cookbook*, a beloved opus in the world of vegetarian cuisine, and several other books. I first met west towards the Pacific, all of the sky form the interior of something, rather than a competing series of exteriors. By making his land so folded in design, he has humanized all he can see.

We arrived a little late and were greeted by Mike's wife, Dianne, whose pleasing grace put us immediately at ease. A row of pear-shaped African gourds three feet high and almost two around sat plumply in the shade of the porch, waiting for Dianne to carve them. She led us out to the cutting shed, where Mike was preparing flowers for an upcoming wedding, and disappeared.

I felt odd explaining our errand to a man who obviously knew more about it than I did, engaged as he was with the brusque transformation of messy, fresh-cut flowers into neat

"There are a lot of us out here—people engaged in guerrilla agriculture."

him in the offices of our mutual publisher, where his gifts of strange-looking, sweet-tasting gourds of exotic heritage earned him the moniker "Bearer of Strange Melons."

Mike Madison haunts order. He built his home low and quiet under a grove of sheltering trees, and all the outbuildings nestle alike under arbors of their own. The layout of fields, cutting shed, toolshed, greenhouse, and home conforms more to a cat's cradle than a simple grid. The alternating rows of tulips, melons, lavender, and Japanese eggplants are straight, true, and weed-free, but rapid changes in the scenery as one walks through the varied plantings on his twenty-two acres, and the sense of new space imparted by rows of bay trees, produce a sensation of cornucopia and pleasant confusion. In the exactitude with which he has read his piece of land and commented upon it, Madison has folded space like a piece of paper, taking a flat earth and making of it a box, a paper lantern. The eucalyptus trees north of us along Putah Creek, the coastal mountains table settings, but Mike didn't seem to hold our ignorance against us. After we explained our ideals, the value of smallness, the tyranny of the supermarket, the corners of the Central Valley dedicated to the refutation of agribusiness, glimpses of a different kind of life, Mike nodded.

"There are a lot of us out here—people engaged in guerrilla agriculture, waging a secret war on the big-ag that seems to be everywhere." Then, striding past handmade racks for sun-drying tomatoes, he took us out to show us the pride of his operation, the melon beds.

Mike walked quickly and nudged melons around with his feet, apparently appraising ripeness through his boots. Two of the three types of melons he planted had failed. September had been unusually cool and the melons hadn't received sufficient sunlight or heat to ripen. He scooped up an emissary of the first crop, cut out an elegant slice from it with a knife I never saw him unsheathe, and handed the white slice to me. Then he did the same for Scott. My thirst drank greedily of the juice gushing from the melon's flesh and I devoured my slice with pleasure and haste. Mike, however, cut himself a transparently thin piece, spat, and tossed the open melon back into the patch without ceremony or a second glance. Chagrined, Scott and I realized that we had just enjoyed a worthless, no-account failure of a melon, a melon undeserving of the name, a melon beneath contempt.

"I can't sell anything from this crop," he said, and hopped nimbly between rows to another variety. The second melon was more promising. Mike dispatched three slices with samurai speed and indifference and a ripe melon scent bullied the air around us. Scott and I reached hurriedly for our share. The flesh was orange and glistening. "Better," Mike grunted, and then dropped the melon and moved on. I lingered some, savoring the sweet fullness, the sensual pliancy. It was not the best melon I had ever eaten, but it compared. This was a melon I would have rhapsodized about had I fetched it for my breakfast table, and yet this man, who was no snob and was proud of his farm's ability to provide for his family, would not, could not sell it. What did he know that I didn't? Did melons actually get that much better?

Mike was cradling a large, strange object and grinning loudly. I broke my reverie and hurried over to see what he had found. It was a fearsome object, and ugly. Its shell was several shades of green and covered with smooth bumps, obviously the victim of some wretched disease. Mike parted the amphibious skin of the melon like Moses parted the waters—it was as if this melon begged for the knife. He told us it was a Piel de Sapo, a Spanish variety whose name translates as "toad skin." Once open, the melon was no longer ugly. Its flesh glistened like melting snow, weeping over its own perfection. The flavor was a cathedral and a liqueur. What did this mean? How had I lived twenty-five years and never been given so sweet a gift? We do not eat

real food, I thought, we do not eat real food. So staggered was I by this mortifying ecstasy that I could not stop Mike from letting this melon—this Aphrodite—drop back into the patch from which it sprung. Scott too, in mute horror, watched the melon fall.

"We'll let the chickens have it. There's nothing they like better than melon seeds."

MIKE HAD BEEN UP SINCE DAWN and at ten o'clock was ready for his morning coffee, but first he took us to see the creek—the salmon weren't running, but he expected them any day. He admired the narrow ribbon of water with the adoration commanded by those things which are ours, but which we do not own and cannot control.

"If you look near the top of the trees, you can see hanging debris. That's how high the water rose last winter."

Staring down at the clear, shallow waters braiding peaceably beneath my feet, it was frightening to strain my eyes upwards thirty feet and realize that standing here at

> Once open, the melon was no longer ugly. Its flesh glistened like melting snow, weeping over its own perfection.

a different time of year would mean that I had drowned. Unsummoned, the immensity of this corner of the world came into a cruel clarity: how Putah Creek leads from the mountains beyond man-made Lake Berryessa to the same typhoon-tortured ocean that crashes against the eastern edge of Japan, the rotating earth spooling sunlight like a ball of yarn, bringing the torrents of winter and the long drouth of spring, summer, and fall, of the water, light, and winds that, before they die, have touched the face of all things...I could really use that cup of coffee, and when Mike turned from his own contemplation and moved back towards the house, I was grateful. I was also excited for a chance to see the Madisons' kitchen, the unacknowledged heart of any home.

> We eat like peasants, whatever's at hand. An hour before dinner, I find myself rummaging in the garden, a few fennels tucked under my arm, my hat full of beans, tugging leeks out of the ground—ought to be able to make something out of that....The characteristic feature of a peasant diet is that today's menu is the same as yesterday's menu is the same as tomorrow's menu.... Monday: tomatoes with basil, grilled aubergine on toast, apricots. Tuesday: tomatoes with basil, grilled aubergine on toast, apricots. Wednesday: tomatoes with basil, grilled aubergine on toast, apricots. Thursday: tomatoes with basil, grilled aubergine on toast, apricots. This changes slowly through the year, so that by October the grilled aubergine has given way to a sardine, and the apricot has been replaced by a persimmon, and maybe there are some new potatoes with mustard seeds alongside.
> —Mike Madison, from *Blithe Tomato*

Being in the Madisons' kitchen, with a diffuse orange light shining through the windows on the rich, amber-bread for dipping. Then he told us about growing quinces. Quinces are related to pears and apples and known for their hard, perfumed flesh. They have an astringency that must be cooked out and have fallen out of favor in this country, but the Central Valley has enough immigrant populations to make quince cultivation profitable. The Madisons have been given recipes originating in Asia, Europe, and South America and have earned themselves the loyalty of the homesick for growing this most ancient of fruits.

"The Mexicans just slice it paper-thin and eat it with salt and lime, though I've never heard of anyone else eating it raw."

Mike is something of an eighteenth-century encyclopedist for whom farming is catalog and exploration. He endlessly varies his crops, searching for hardiness, adaptability, and cycles of ripeness that can create year-round cash flow, and popularity amongst his clientele. His experiments are, as a rule, successful, but he let us in on one of his failures. A few years ago, he scoured the world for table grape varieties and picked out a handful

---

Truly great food, like truly great sex, exists. But neither is discovered without some intimacy.

---

colored wooden counters and the copper pots and pans hanging on the walls, is a bit like visiting the inside of a jar of marmalade. Without the influence of nature to complicate his schemes of organization, Mike constructed his kitchen as an essay on the pleasures of reason. Cutting boards, utensils, baskets of fruits and vegetables, and a few cookbooks—including one of his sister's—are all within two steps of each other, but also so arranged that four people could be at work in the space with only the most pleasant amount of bustle to impede them. We each sat on stools around the kitchen island. Mike poured some of his own small-press olive oil into a dish and cut up some known for rare flavors and sweetness. No one around was growing these grapes, and he looked forward to his first harvest.

"The fruit was beautiful, but when anyone tasted them and found the seeds, they wouldn't buy them. I had to tear all the vines out." He spoke without bitterness or even sentiment, but with a scientist's puzzlement. The grapes were so perfect, in theory...it really should have worked.

We'd been talking in his kitchen for more than an hour, and Scott and I both felt that he was anxious to return to his long day's labor. Citing our gratitude for his

PLATE 14

generosity, erudition, olive oil, and coffee, we began to take our leave.

"Well, you have to take some figs before you go. A couple trees grow feral on our property."

Feral figs?

We followed Mike out to another corner of his place and found two shaggy, wild-looking trees thick with purple, tear-shaped fruit. Scott and I immediately set to work harvesting our untamed bounty. The figs were so ripe that the rustling of our activity caused much fruit to fall to the ground, and I scooped these up as well. When we had acquired more than twice the number of figs we could possibly eat before they turned on us, we stopped. Mike asked where we were headed next, and we made some vague answers about roadside stands.

"You should go see Ramón, another farmer that goes to Davis. His place looks like a weed patch, but every week he brings in beautiful peppers, eggplants, onions, and squash. He's got quite a story."

We promised we would and headed down the road in that direction. On the very first visit of our very first day in the Valley, a set of scales had already fallen from my eyes. In many food books akin to the one I hoped to write, I had read, yes, ad nauseam, of the orgiastic delights granted to devotees of ideal produce, but I had taken these descriptions as harmless hyperbole, a bit of showbiz. I was wrong. My first Piel de Sapo, lying discarded in a field full of melons just like it, was a road to Damascus. Truly great food, like truly great sex, exists. But neither is discovered without some intimacy. If I looked for these edenic pleasures in the supermarket, I'd never find them. This was a tiresome discovery. It meant I had to begin living differently.

Halfway to Winters, Scott and I had already devoured roughly fifteen figs each and still not worked one-third of the way through the box.

"I like these figs," Scott said. "They're completely ripe but not filled with spiders." And just as he said it, I bit into the wrong one. Funny, the suddenness with which knowledge moves.

Wherever People Are Hungry

Scott and I were traveling through the Valley with overlapping but distinct spheres of interest. At first, we didn't even have a name for what caused us to stop the car and walk about. We defined our quarry in terms of exclusion. Smallness seemed to be the key to everything. The soft tenets of edibilism were only discernible in the dissatisfaction I felt with the current debates surrounding food and ethics. I sought an ethos of eating that could give rural America back to itself, and, as a consequence, break the bonds of global predation that diminish rural places worldwide. But I am not dogmatic, nor expert, and I lack both the talent for research and the patience for sustained argument. I am just a young man from the Smoky Hills of Kansas nursing a quiet vengeance in his heart. So instead of facts and figures and the famous, we sought people we didn't know until we found them, and whatever truth inheres to life as it's seen by accident.

Scott's calling, on the other hand, was as honest as the Rolleiflex camera he carried. The Central Valley has for decades been a curious and provocative subject for photographers, and every time he passed through it those photographs and future visions of his own obsessed him. He had photographed Cairo, Budapest, and the Dakotas, and when I suggested a modest three-day survey of roadside stands in the Sacramento Valley, his enthusiasm was immediate. We still did not know the magnitude of our project as we followed our scrap-paper directions to the home of Ramón and Lucy Cadena.

Putah Creek wends north and east towards the Vaca Range, loses its modest singularity in the broad waters of man-made Lake Berryessa, and then regains it in the peaks beyond. The lake heaves and billows in a range of shallow mountains, but it is served by a nearby valley town: Winters, California. It was here that we first noticed the extreme permeability of the Central Valley. The various boating and fishing outings inspired by the nearby lake support innumerable bait and sporting goods shops, which add a dimension of middling wealth and mild festival along the edges of the municipality. With San Francisco but a forty-minute drive away, enough traveling cosmopolitans with bulging wallets and heavy purses pass through to help sustain a boutique winery with tasting room, the award-winning Buckhorn Steak & Roadhouse and the Palms Playhouse, an old opera house converted into an intimate concert space featuring largely blues, folk, and traditional country music. This arrangement is beneficial for townspeople, Valley travelers, and coastal day-trippers alike and has created a charming if slightly discordant town.

Plate 1.5

We turned north from Winters and headed up a lonely road to desolate Esparto, took a left at the abandoned post office, and were soon pulling into the driveway of a two-story farmhouse in the porch-swing style. We found Ramón and Lucy Cadena sitting in the shade of their toolshed, shelling walnuts with slow, steady hands. Buckets of walnuts sat at their feet, picked early that morning. Ramón and Lucy had begun shelling at eight and, had we not interrupted them, would have shelled until mid-afternoon, gaining a whole bucket of walnut meat for their trouble. I have never seen a young man or woman shell walnuts by hand. Old man Muir would conscript me every fall on pretext of my family's poverty, and I would grudgingly sit in his basement, helping him crack the hard round shells and

gardens created a much more familiar pattern. Roses of all hues and sizes had been pruned and shaped with love, and the eye was drawn not to the broad brushstrokes of Ramón's farming, but the gardener's pointillist obsession with detail. The line between the two worlds was sharp and definite, but without any discernible animosity or signs of border skirmish. All the treaties held.

We apologized for interrupting him, but Ramón said he was happy for an excuse to quit, and Lucy, also pleased to be freed from the task, left her half-shelled bucket of walnuts and went into the house.

"See those peach trees over there?" Ramón said, pointing towards a short double row of fruit trees, now out of season. "These are my wife's trees. She loves peaches,

Ramón's work was a verdant chaos of edible splendor—kale and chard and basil mixed in with onions, peppers, and tomatoes.

extract the ear-shaped meats of the truckload of nuts he had gathered from neighborhood and countryside trees. My payment for shelling walnuts was never anything but walnuts, and one year I lost a baby tooth on a hidden piece of shell. As long as walnuts can be shelled by hand, I suppose they will be.

Ramón and Lucy's parcel of land was only just over an acre, little more than a cultivated lawn, but shaped by two hands. Ramón's work was a verdant chaos of edible splendor—kale and chard and basil mixed in with onions, peppers, and tomatoes, and all surrounded by and intermingled with various weeds. The sheds were a mess, with tools perpetually laid down randomly after use and abandoned or half-finished projects. Bits of cars and engines were overgrown by golden grass along the edges of his buildings. This abundant disorder ceased abruptly near Lucy's house, from whence radiated a shock of floral cleanliness. Here, trimmed green grass and intricate

and whenever she tastes a perfect peach, she saves the seed and I plant it. All of those trees are different."

Ramón's farm was a bit like an Easter egg hunt, or a game of hide–and-seek. Each fruit and vegetable waited not only to be observed, but discovered. Several kinds of winter squash, red onions, hot peppers, sweet peppers, dull green and purple and forest-colored greens, eggplants, and tomatoes appeared as Ramón knelt and waved weeds and overgrowth away from his produce.

"These," Ramón said, holding a distinctively curled pair of peppers upside down, "are called 'horns of the bull.'"

"See this?" Ramón asked, pointing to a waist-high stalk with wide, spade-shaped leaves and no visible bud, berry, or blossom. "This is Mexican wild yam. You eat the roots. It grew on the riverbanks where I lived with my brothers as a kid. Now they found that it's healthy for you. They call it Mexican ginseng. I have been remembering it for years, but I just now got some seeds. In a month, when I dig it up, there should be thirty or forty little yams under this plant."

"What does it taste like?" I asked.

"Well, William, you'll just have to come back and taste it for yourself." I told him that I might just do that, and we continued the hunt. Ramón pronounced his words with easy care, with a seductive tone that hinted always and without urgency at unspoken wisdom. It lulled me, not into sleepiness but into a slower kind of wakefulness. Smoking a pipe on my front step as I stare at passing clouds produces the same effect.

Scott and I play a game in the Valley, a reading game. Whether on the road or walking through a friendly stretch of land, our eyes interrogate the earth and its inhabitants for coded information. What has been touched, and why, or left conspicuously untouched? What is the scale of things, what has been touched by machines or by hand, what by an accident? This game, the single rule of which is

> "It bothers me, all these oil problems, and this development. Where do the people of this country think they'll get their food?"

that the landscape is legible, has revealed much to us and directed us to the people, farms, ranches, and questions that made this journey possible.

On the edge of Ramón's acre grew a kind of living fence of broad, green paddle cactus. He had planted cactus, I thought, because it can and will grow without water and is at worst a weed, at best decorative. However, the branches reached uneven heights and the top and sides of each paddle were heavily scarred—clear evidence of human knife work, of harvest.

"What's this?" I asked.

"I'll show you," Ramón said, and unfolded his pocketknife. He carefully inspected the paddles, explaining his actions as he went along. "This is too big," he'd say, or "This is too small. What you want is these soft, bright green paddles, like this one." Ramón cut off a pliant paddle at its base, deftly shaved the furlike thorns off the face and edges of the plant, and cut Scott and me each a thin, glistening slice.

"They're just like green beans," he said, and Scott and I put the oozing, gelatinous slices in our mouths and began to chew and make noises. They were nothing like green beans, but they were refreshingly sour, and we both reached for another slice. The curiosity was pleasant, and the taste and texture not unpleasant: more data was needed. I began to get a sense for their virtues, and imagined them sautéed quick and hot with sliced serrano peppers and strips of skirt steak, or puréed and added to a soup of pinto beans and ham hock.

"What do you make with them?" Scott asked, on a similar train of thought. Ramón seemed a bit lost by the question and talked about omelets and frying them in pans with the vagueness of a man who has been banned from the kitchen for several decades.

This was our first encounter with the half-wild, half-cultivated nopales cactus. We would meet it again, consorting with rosebushes in an empty lot in Isleton, being carefully poached along Highway 5, and growing in rows on a small lot on Kings Canyon Road east of Fresno.

"Where does it grow?" I asked, wondering rather forlornly if it could survive a Kansas winter and ugly up a couple of hedge-appled ditches on my land back home, but Ramón's answer surprised me.

"It grows," he said with unwavering eyes on mine, "wherever people are hungry."

I broke Ramón's gaze and looked again at the scarred desert fence, and saw, through Ramón's eyes, the manifestation of a benevolent God. I was reminded, with

PLATE 1.6

PLATE 1.7

new devastation, of Rilke looking at the broken torso of Apollo and seeing through its dismemberment to the god himself, and writing:

Otherwise this stone would seem defaced
beneath the translucent cascade of the shoulders
and would not glisten like a wild beast's fur:

would not, from all the borders of itself,
burst like a star: for here there is no place
that does not see you. You must change your life.

and with all the energy endemic to blasphemous youth, I changed the subject.

"You don't use any pesticides, do you Ramón?"

"No, William, I don't."

"Do you have any trouble with insects or gophers?"

"I'll tell you, William, what I do. I always keep weeds on the edges of my plantings, so that everyone has enough to eat. And then I say a prayer to St. Francis that these poor creatures will not be so hungry that they must steal from me," Ramón said, his hands clasped behind his back, a nopales-wet knife still drawn, and his head bowed in an attitude of deference and trust.

On our way back to the rose-hedged patio where Lucy would serve us coffee, we passed a shady enclave that Ramón used to store his winter vegetables, and judging by the reclined lawn chair, used also for his own midday naps. I pondered the strange providence of this fall harvest. Squash and onions are heavy, ursine sleepers. But it is their slumber on which we feast when the earth no longer bears its green burden; they dream so we may not starve. They are unlike their summer cousins: peaches, strawberries, and tomatoes hurtling towards rot, hosting insect hordes, leading brief grasshopper lives. Among his winter vegetables Ramón might be overcome with languor and lie down, letting his eyelids fall among the mingling of red onion skins and autumn leaves. A discarded soda can, even, did not lessen the beauty of the scene.

Lucy brought out a tidy tray carrying cups of coffee and a honey bear. The bottom quarter of the bear was honey, the rest contained cream. Ramón asked her to sit with us and began to tell his story.

It begins on the streets of Guadalajara, where Ramón and his brothers roamed as orphans. Ramón began working in the fields for pesos a day and lived off food he could gather and catch himself. Through this hardscrabble and feral existence, he came to be a young man and started looking more deeply into the world around him. One day, on his way to the fields, he saw several old women gleaning food from the edges of the fields and herbs from the ditches. When their baskets were full, they went to the city and sold their goods directly to citizens on the streets, thus making many times the money he was being paid for his rise-to-set days in the fields. So, enterprising young Ramón got a basket and started gathering and selling whatever he could glean himself. Being younger, he could make several trips from the country to town to sell every day. This lasted a lucrative week or two before he was set upon by a mob of old widows who accused him, a strapping young man suited for real work, of stealing bread out of their helpless mouths. So he had to give up his new career. Shortly after, he met and began courting Lucy.

Lucy had been listening restlessly to the story so far, breaking in occasionally to tell Ramón that he talked too much and fussing over his pant legs, which she pulled lower towards his shoes, and his socks, which she pulled higher up his calf, but now she spoke with the kind of genuine affection that requires a mask of annoyance.

"Oh, he was an impossible boy. He wouldn't leave me alone. He would come to our street and shout up at our window, and I would tell him to go away. Then he would go away but come back the next day! I'd never met such a man, and I stopped coming to the window. I sent my mom to tell him to go away, to tell him that I didn't like him at all. He said that was fine, and that he'd be back tomorrow."

Ramón interrupted her and finished the story of the courtship with sudden economy. "And so we got engaged and I went to work in America through the bracero program so that I would have enough money to support my wife."

He worked a summer season and then his employer asked him to stay that winter. That winter, alone in a bunkhouse, looking for something to read to ease his loneliness, he found a matchbook advertising a program for learning English. He had some money saved, so he sent away for the books and began to learn English. Ramón continued to be valuable to his employer, who

stones around each one," Ramón continued, and put a hand on his wife's leg.

But years ago Ramón was badly injured and could not work as much, and so he couldn't afford to drive from that house to work every day. He had to move closer, and they sold the house he had built and bought the one they live in today.

"It bothers me," Ramón began, changing tones, "all these oil problems, and this development. Where do the people of this country think they'll get their food? I know that as long as we're here, on this land, that we'll never starve. I don't know what others are going to do."

I realized then how much hunger informed Ramón's relationship to the land. Hunger, for Ramón, was a very

---

It was a compelling and sobering understanding of agriculture, that the farmer should seek to feed himself and then the world, not the other way around.

---

sponsored him for citizenship, and he began to build an adobe brick house for his wife, in the Capay Valley.

"How did you learn how to build an adobe brick house?" Scott asked.

"I'll tell you, Scott. One day, in Mexico, I saw a man building a house out of bricks all by himself and I shouted at him that I wanted to learn how to do that. He said he would teach me if I bought him a bottle of whiskey. So I went out and bought some whiskey, and he drank the whiskey and taught me."

"Oh," Lucy said, rubbing suddenly teary eyes, "don't talk about that house. It was so beautiful. I first saw it and knew it was the most beautiful house I had ever seen. I still dream of that house."

"I planted fruit trees all around it, one of every kind, for her. I painted the trunks white and put a ring of

real evil. He did not talk about going hungry, but an orphan on the streets must have done without more than he did with, and Ramón never forgot the feeling; he didn't even wish it on the insects and animals that might nibble his fruits and vegetables. He worked the land, after all, not only to feed himself and his family, but the people who shopped at the Davis Farmers' Market as well. In hard times, he could share with the bugs if they needed the food. His belief seemed to be that everyone should farm so that no one should starve. It was a compelling and sobering understanding of agriculture, that the farmer should seek to feed himself and then the world, not the other way around. Hunger lives, in Ramón's mind, just days away from all of us without well-worked land and could strike at any time. How could one live with such insecurity?

"Have you seen our pomegranate trees?" Ramón asked. "They are right out front. They're over one hundred years old. You can tell because they are different from any pomegranates you've ever had. They don't plant them anymore. The seeds are sweeter than the others." So I went out to harvest my own ancient pomegranates, found a handful in the shaggy tall trees, juggled the fruit while walking up and down the empty dirt road, and then took the Cadenas' mail into the kitchen while Scott made a final sweep around the acre, looking for pictures.

As we were leaving, Lucy asked us more about our book, listened with interest, and then told us with a serious and giving voice that she would pray for the success of our book and for safe travels. We bowed our heads and gave her our gratitude.

As we drove back to Winters we talked about Ramón and Mike, their very different yet somehow mutually sympathetic methods of farming, the solidity each possessed, and their relationships to their land and how it provided for them. It became clear, then, that our purpose in the Valley was somewhat more than being spectators. Once we began to know the people and the intimate corners of this land, grand, vague ideas began to take shape.

We decided to create a portrait of the Valley composed not of its dominant landscapes and activities, but only of those that bore a human, rather than mechanized, imprint. A collage of moments, the intersection of time and place and people—irreproducible, but alive in ways that anonymous portraiture of a line of field hands is not.

PLATE 1.8

Famous Everywhere but Brentwood

"DID YOU SEE THAT?" Scott asked excitedly. The sign was a hand-painted muddle of colors, vaguely vegetable in aspect. Over this backdrop ran the letters TAIRWA. It stood next to a driveway on a small, verdant patch of land.

"Yeah, but we're not yet in the Valley," I said, continuing to hurtle forward.

Scott studied the flat landscape of well-worked fields for a moment, looked back at Mount Diablo in the distance, and asked me what the hell I meant.

"We haven't crossed a bridge yet—there's one just a few miles up the road; you can only get into the Valley if you cross a bridge," I said, thinking the truism true enough.

Scott pulled out the map to check my assumption. "Assume: an ass out of you and me," my grandpa always said, and he was right.

"Fine," I retorted as I turned around. We'd only been on the road for half an hour and had just passed through Brentwood, a small city in the desperate paroxysm of meaningless development, a lost place. It just seemed too soon to stop. I could still feel the shadow of the bay.

"Tairwa" is a fast and loose phonetic rendition of the oenophile's *terroir*, a term used to indicate the unique elements of soil composition and microclimates that theoretically distinguish one vineyard from another. It is the most metaphysical of all the wine conceits, which is not to say I dismiss it, and Knolls Farm uses it for good reason. The farm of Rick and Kristie Knoll not only conforms to organic standards but has gone a step further, into a philosophy called biodynamics, first developed by Austrian scientist Rudolph Steiner in 1924. Basically, a biodynamic farm imitates nature by becoming a wholly integrated ecosystem of which the farmer is also a part. Insects, plants, and animals should all aid one another from birth to death, and, ideally, no outside nutrients should be needed. Biodynamic farmers have also developed various systems of microbial compost to aid in the absorption of nutrients and fight off disease.

"It's a good thing we didn't use the French spelling. After September 11, everyone would have thought we were selling terror," Rick mused.

The Knolls live in a small house set in the middle of their ten acres. We walked past beehives, wary chickens, rows of fruit trees with rosemary, oregano, mint, green garlic, arugula, and fava beans planted in between—a serious departure from nearly all the orchard land we'd seen where the ground was dry and flat to aid a mechanized harvest. Save for the driveway on which we walked, there didn't seem to be a single patch of exposed earth on the whole place. Next to their

PLATE 1.9

front porch a small pig squealed but showed no interest in my attentions. Fine, pig. See if I care.

Kristie handles most of the business and Rick does most of the actual farming. They also employ a handful of workers, all Latino, including a full-time delivery driver. Kristie and Rick sell everything on their farm either to organic grocery stores or directly to restaurants, so there are a lot of deliveries to be made.

"We used to go to farmers' markets when we first started out," Kristie told us, "but we didn't like all the…"

"People," Rick finished for her.

We could see his point. Kristie seemed at ease with us, but Rick, with his reptilian movement, angry intelligence, and strange red rims around the lids of his eyes, gave

off green things and putting them in his mouth, suggesting we do the same. Who knew? Bean and pea greens are delicious.

"I don't even consider myself a farmer anymore. I'm just part of the ecosystem. My only job is not to fuck it up."

I liked Rick's idea, but looking over at the hunched figures half buried in row after row of green labor, I wondered if he would feel that way if he did all of the work. He might not think of himself as a farmer, but I bet the women snipping and bundling flowering rosemary thought of themselves as farm workers, albeit in a more pleasing setting and with more varied chores than usual.

A gap through the tree line that makes up the Knolls' property line revealed a barren moonscape of maybe one

---

Save for the driveway on which we walked, there didn't seem to be a single patch of exposed earth on the whole place.

---

off the impression that he really had more use for his microbes than he did for larger, more clumsy animals like human beings, a worldview he might have acquired along with his PhD in chemistry. For both of these refugees from suburban Santa Ana, one of the great appeals of rural life must have been the solitude.

"Where did you hear of us?" Kristie asked and was surprised when we said we had just driven past and seen the sign.

"This is one of the first organic farms in California. We're really well known. We're famous…well, worldwide."

"Famous everywhere but Brentwood," Rick said.

Kristie showed us last year's product list, three dense pages featuring four kinds of figs, five kinds of apricots, nectarines, plums, flowers, herbs, greens, tubers, and strange terms like "amaranth," "cardoon," and "lamb's quarter," and then Rick led us around the property, picking

hundred acres that had just been disked.

"Look at that," Rick said with disgust. "An ecological holocaust."

Rick left us for his work and let us wander about the ten acres on our own. We waded hip-deep in thick rosemary bushes where, Kristie had told us, coyotes make dens and pup, we chased feral chickens who could fly to the top of the high eucalyptus trees, and we climbed orchard ladders near fruit trees, tasted and spat out an uncured olive, and let the invigorating abundance of the land get in between our toes and in our hair.

Our methods of inquiry, as photographer and writer, had become intricate and complementary. We didn't so much interview people as chat them up a bit, and Scott, being more comfortable in his skin and possessing a journalist's background, generally spoke more at the beginnings of our encounters. My role was to overcome my inward tendencies and take over the conversation so that Scott would be free to seek his photographs. Over time, I came to intuit something of Scott's vision, his appreciation for the time-marked material, for tools, and knives, and the presence of the ground beneath our feet. Color seems to inhabit all five of his senses, a dimension of its own. Most impressive, to me, was the empathy he brought to his subjects. Rather than distort his images, this reflexive empathy has made them more present, far-seeing, and immediate. His approach is more omnivorous than predatory; more the nose's question: *What's this?* than the stalking, chasing, and pinning that characterize much photography. After watching him work, learning a couple elementary principles of lighting, and discerning, out of the range of good photographs, which were most distinctively his, I was able to enter into that world and ask him "what's this" myself.

"Hey Scott," I said, finding him looking askance at a young German shepherd gnawing at a shovel that had been stuck in a tree. "You've got to see this"—great blue piles of cut rosemary.

THE KNOLLS HOST PICNICS for their customers period-ically. Bay Area foodies descend on the ten acres, bring wine and blankets, eat, socialize, and revel in the horticultural beauty and their cleverness for knowing about this place only a half-hour away from their houses in the hills. It must be a wonderful time, but the thought prodded something bitter in me and I felt the need to move on. I had been right all along. We weren't in the Valley.

Suddenly the nigh miraculous productivity of the land, the year-round harvest, and the enormous variety seemed almost sinister. There was no doubt that the biodynamic approach was more productive than so-called conventional farming, and that such care was far better for the land than mining it or poisoning it, but here, it seemed, the land's natural productivity had been made unnatural. Nature knows abundance, but it also knows rest. That was what was missing from this landscape—leisure that didn't rely on the paid labor of others. I'm not condemning agricultural labor—that would hardly do anyone any good—but when Rudolf Steiner advocated for ecological harmony and self-sufficiency in the working of a piece of land, I doubt he meant for the industrial machine to be replaced with a human one.

Jeez, commented one of the saner voices in my head, *there's just no pleasing some people.*

My purpose in these travels is not, however, to be content. I'm looking for an aggregate of knowledge that I might carry with me into another life.

Through their passionate and articulate opposition to poison-based agriculture, the organic movement, or at least the terminology it invented, has now thoroughly succeeded in the marketplace; indeed, the demand is now more than true organic farming can provide. Now that the government has created its own organic standards, many of the movement's pioneers have grown disgusted with the corruption of their ideals (not to mention the mainstream competition for their niche markets) and have found new ways of describing the purity and goodness of their produce. Rick and Kristie have adopted the biodynamic label and have made the post-organic transition fairly painlessly, owing to their long relationship with most of the restaurants and markets for which they provide.

My own dissatisfaction with the organic movement is more implacable. By limiting its critique of modern

PLATE 1.10

agriculture to the use of pesticides and chemical fertilizers, the organic movement failed to address either the annihilating consequences industrial agriculture has on rural life or the inability of the lower classes to afford organic food. Part of the reason for the movement's success might lie precisely in this limitation. If farming is an environmental rather than social issue, the role of the consumer is enhanced. The consumer who buys an organic peach is assured not only that it isn't cloaked in invisible poisons, but that he is in a position to change the world for the better. Which is a nice feeling, after all, to be able to afford. If only, the organic consumer believes and has been told, everyone were like him.

Ultimately, the organic movement belongs more to the history of marketing than to agrarianism. That certain farmers have found a way to treat their land well by providing first-rate goods to wealthy people is admirable business but is not especially moral. Consumers that don't want to give themselves cancer are simply amoral. The master's tools, in this case consumerism, will not tear down the master's house, which is industrialism.

We require more farmers, and a thriving, diverse, and autonomous rural culture to support them. We need, and need to become, the makers of things.

"That was nice," Scott said as we pulled away, "but we need to get deeper into the Valley."

I Am a Jam Maker

"We don't charge you not to poison you," Paul Buxman said when I caught him on the phone. "I don't think that is behavior that should be rewarded, it should be expected. I'm a human being. You're a human being. If you eat what I grow, you won't be poisoned, but I won't blackmail you for it. I won't extort more money from you. Because if that's the system, what happens? The poor can't afford food that will not make them and their children sick. And if I'm the one demanding prices they can't afford, then I'm effectively withholding health from them. Why would I do that? For more money? We believe in a nonviolent approach to farming. I began in the organic movement but quickly realized that it wasn't enough not to poison people. I had to go farther. What does nonviolent farming mean? It means taking care of the land, growing healthy produce, but it also means taking care of the people around you—being involved. Nonviolence does not avoid violence, it gets in the way of violence. I learned that from my wife, Ruth. She's a Mennonite minister, and she's traveled the world to protect victims from violence. So whenever someone works for us, which isn't often, they may live with us, we pay them a living wage and get them what benefits we can, and we sell the food at the right price. This means I may never own my farm, but what do I need to own it for? I can work it, make it productive, and live off it. We're involved in the community. This year we took a group of, well, let's say 'at risk' youth out to the farm and had them plant melons. They weeded, watered, and harvested the melons. We put their pictures up in some of the places that sell our produce. It was really beautiful to see the kids out here, learning something about farming, learning about another way of life than what they know. We paid them for their work and they kept the money from the melons. It was a great program. But what is it, William, that you wanted to know?"

"Well, this all sounds great, Paul, but what we'd really like..."

"I tell you what, William. What you really need to do is come down and visit. Spend the day with us and we'll show you what we mean," he said, obviously loading his cannon for another barrage of moral argument.

"Great, Paul, we'd love to come down. Next Sunday after church?" I asked with some urgency.

"Sounds great, just give us a call."

I put down the phone, excited. So much of what Paul said resonated with me. *Nonviolent farming*. Unlike much of the organic movement, here was a philosophy of farming that gave human and natural concerns equal consideration. The Buxmans started a movement called California Clean, which requires of its members not only environmentally sound

Plate 1.11

farming, but socially responsible practices. A California Clean farm should not be any larger than a family can usually work on its own. A Clean farmer must pay a living wage to anyone who works for him. A Clean farmer must not inflate the prices of his produce. That's a serious set of demands for someone working the land in the United States, where a system of government subsidies and economies of scale has forced food prices to remain very low, and where land prices and property taxes are high, but it is also a philosophy with the capacity to change that system. Unlike the organics. South Africa, Australia, and New Zealand have all begun their own Clean movements, which, possibly because the organic movement is not so entrenched there, are thriving.

WE STOPPED IN KINGSBURG, a small Swedish community so desperate to avoid the soulless Fresno sprawl that it pumps noxious folk music through its empty downtown via pole-mounted loudspeakers in a sad attempt to simulate real culture, and then headed east towards the purple Sierra Nevada. The hour was late and the air was reddening like chafed skin. It had been a long, sweaty drive in a Volkswagen without an air conditioner and with a temperamental CD player.

The road was soon surrounded by orchard and the fine Buxman home came into view. Ruth met us at the door and said that Paul was on the phone. We looked at the fading light and hoped he wouldn't be too long.

Ruth told us something of Paul's history. It turns out that Paul Buxman is, in a small but life-altering way, famous.

His story is a dramatic one. A second-generation conventional farmer, Paul renounced the herbicide and pesticide model after his young son developed cancer from exposure to the poisons. Paul's revelation came at a crucial media moment when other, similar stories had just begun to surface. When the television newspeople saw Paul, a tall, boyishly handsome man bursting with sincere outrage, sorrow, and pristine rhetoric, they knew they had found a treasure, and every few years since that time, when his story becomes relevant again, they descend on his little farm with their vans, microphones, and cameras.

Paul emerged just as the sun vanished and took us in the near dark around his place.

"The apricots, peaches, plums, and pluots are all done now, but we still have some watermelons that the kids planted," Paul said, and then he started in on a speech he must have given a hundred times, about the seasons, how the soil is a living organism, what it means to be a farmer, and why the culture of poisons is loathsome and doomed to collapse.

The watermelon patch is hidden by an organic fence. A row of corn and tangle beans protect it from sight. The corn hides the melons, and the beans that connect stalk to

---

"I used to think I had figured out one hundred things about farming."

---

stalk make breaking though the fence loud and difficult. Paul grabbed a Black Beauty watermelon for us, cut it in half, and tasted it. The flavor was not what he wanted, and so he turned the red face of each half into the dirt.

"Thieves are like magpies," Paul said. "They are drawn towards the color red."

He searched the field for a proper melon. When he found one, he left it on the ground and bent over it with his knife.

"When a watermelon is really ripe, it will just split apart for you. Watch."

Paul slid just the tip of his pocketknife into the top of the melon. A moment later, we heard a loud pop. The melon cracked right in half and then fell open, revealing flesh that was the deep red of a beating heart. Scott and I were staggered. The heavy ripeness of the watermelon had actually ripped it apart.

"We'll take this one inside."

As we walked back to the house, Paul told us about the loco ant and the antlion.

"The loco ant is named that because of the way it walks. Its bite feels like a thousand wasp stings. Have you ever seen an antlion? No? It builds these traps—see, there's one

winning painter of bright Valley landscapes, and in the heart of the house, the spacious living room, every wall is covered with Paul's colorful, peaceful work.

"Have you ever heard of Guinness?" Paul asked. My heart bellowed like a young calf. A Guinness sounded wonderful.

"We know each other rather intimately," Scott said.

"Well, you know how you can't see through it? I make a grape juice just as dark. No light gets through this juice." Paul went into his kitchen and returned with two brimming glasses. My disappointment lasted only until I tasted the juice. It was completely dark, but just along the

---

"Only a few fruits are right for jam. I look for the largest, most ripe apricot or peach I can find....When I pick it, it should ooze all over my hand. Then I know it is ready."

---

now—and when an ant falls into it, the antlion explodes out of its lair and grabs the ant. There's no escape. Here, let me show you."

Paul coaxed one of the roaming ants onto a stick and dropped it in the sand funnel of the antlion. The ant struggled to crawl out, but the loose sand gave way from its legs, and the ant slid over and over into the bottom. A minute passed.

"I'm really disappointed in this antlion," Paul said, and freed the captive.

PAUL BUXMAN BUILT HIS OWN HOME, making him the third farmer, after Mike Madison and Ramón Cadena, we'd met to do so. The house is two stories, broad and tall, made of thin slats of naturally stained wood. The natural wood look continues on the inside. Paul is also an award-

edge of the liquid it turned a translucent blue.

"It looks more like watercolor than juice," Scott said.

A grape juice had never been as sweet, as tart, or as crisp as this. Grapes themselves, at the peak of their ripeness, only hinted at the flavor of Paul's wonderful liquid. Moreover, it cooled me as it swirled from my lips into my mouth and to my throat and then rested in my stomach. My head cleared of all fog for a moment. My eyes grew sharper. All heaviness of the day's hot travel drifted away like smoke. Had I been drugged?

"I steam-juice the grapes," Paul explained. "I use this wonderful Finnish juicer. I set some grapes in, let the steam work on them, and two hours later, the grapes look like clear pieces of plastic. All the color and all the flavor have been drained into the juice. And that's where all the nutrients and antioxidants are—in the color. This juice can work miracles."

PLATE 1.12

PLATE 1.13

Paul only had a few Concord vines, so the juice was purely for family and guests. A part of me doesn't believe a juice like that should be sold. It lives in a world above money.

Slowly nursing our precious nectar, we followed Paul out to the refurbished shed where he makes jam.

"Only a few fruits are right for jam. I look for the largest, most ripe apricot or peach I can find. Usually it's on the very top of the tree, where it has had the most sun. When I pick it, it should ooze all over my hand. Then I know it is ready. The rest of the fruit goes to the store. I believe that jam should be as close to that ripe fruit as possible. So we cut up and flash freeze the fruit and then keep them at twenty below until we are ready to make them into jam. We use as little sugar as possible; this fruit doesn't need much. My wife, my son, and I all get into this kitchen and just crank out the jam, quoting our favorite movies, singing songs. We just have a great time."

Paul showed us his glowing freezers of flash frozen fruit, gave us a couple jars of jam, and led us back to the house.

"See that?" he asked, pointing to a glowing heart made of little red lights hung on the side of his house. "I keep that lit up all year. Because that's what this is all about. Love."

Back inside, Paul told us a little about how his thoughts on farming have changed. Once a purveyor of knowledge, he now communicates his experience of farming through tone and, more and more, simply being.

"I used to think I had figured out one hundred things about farming. I would go around speaking, telling people of these hundred things. Then I thought maybe I knew fifty things. Then twenty. Then I thought I knew five solid, real, unchanging things about farming. Now...now I've had to stop speaking. I turn down all my invitations. I tell them, 'I don't know anything about farming.' If someone asks if I'm a farmer, I can't say. I don't say that I am a painter. Am I a jam maker? Yes. I am a jam maker."

It was late now, and Paul wanted us to stay. He makes heartbreaking biscuits every morning and eats them with butter and jam, but we turned him down. I'd been three days on the road without a change of clothes. I missed my Olga, I missed my cat, and I'd forgotten my toothbrush. Paul loaded us up with jam, postcards of his paintings, glowing watermelon, a Mason jar of juice, and sunflower seeds. He showed us where Saturn and Mars were in the winking sky. Then we were on our way home.

PLATE 1.14

Roadside Stands: A Pastiche

In a time and place inconsequential to human history, a herd of goats became habituated to taking a certain trail from glade to glade, from the gushing mountain spring, through the blackberry bushes, up the cliff where the clumsy-foot wolves could not follow, and finally to the hidden valley filled with clover and ancestral dens. The earth, under the tread of fifty animals, each shod with sharp, hard hooves, compacted and refused to accommodate vegetation any longer. A path was born.

Roads are, from the beginning to the present day, an expression of herd animals searching for and finding food. Predators may use roads in order to track prey, but they do not require or create them. The footfalls of an omnivore are too distracted to make a road; bears, raccoons, and rats thrive on chaos and variety. But the herd requires stability, and so, as a testament to the faith in a cyclical, unchanging universe without which civilization is impossible (and why, perhaps, it is doomed), goats invented the road.

The roadside stand is a result of the crisis of harvest: in the fragile summer fruits, ripeness is controlled rot, but that control is tenuous and short-lived. The city market is too far away, and the farmer cannot spare the time to truck the bounty far off the farm. The peaches must come off the trees, the melons off their vines. So the waste-hating farmer takes his ripest fruit to the moving city—the road. And the herd pauses there. And grazes.

We visited dozens of roadside stands in the year's worth of trips that we spent in the Valley, and each showed us a different aspect of what at first had seemed like such a simple encounter between grower and eater. As each stand embodied the personality and history of those running it, the wide and pleasing variety should not have been too surprising.

Bruce and Willa Gossett have been running a stand in East Nicholas, a ghost town north of Sacramento, for over thirty years. They store their produce in a defunct cooler for a forgotten local beer called Tavern Pale. Their son, David, began working the stand a couple days a week after he lost his arm in the Sacramento shipyards.

The design of their stand is typical. Bruce or David lounges behind a single wooden counter shaded by a bit of roof that lowers to close the stand in the evening. An old scale holds a knife and an apple sliced for sampling. The produce

Plate 1.15

mostly comes from the Gossets' small farm, but they buy some melons and nuts that they don't grow from other farmers, as well as honey from Marysville native Harold Dirks, who we'll meet later.

Highway 70 between Sacramento and Marysville roars with traffic, and I asked David if the increased population had improved business.

"Nope. Made it worse. It's dangerous to stop now. You could get killed."

Business has steadily decreased, according to Bruce, since the seventies.

"Our average sale used to be around thirty dollars, and sometimes up to a hundred. Why? Because everyone was canning sauces, making jams, pies, preserves, and pickles. That used to be just as much a part of the harvest as fresh

The second type of roadside stand is not connected to a farm, but buys from local farmers and usually runs year-round. These act as small, independent grocers for the community and hungry travelers.

The third type of roadside stand is an outgrowth of the grotesque excess of the Valley's big agriculture. This corporate stand is either an enormous to-do with flags and an eighteen-wheeler full of produce, local and imported, usually no better and sometimes worse than you would find in a grocery store—or it is a much lonelier scene. Vans will drop off lone migrant workers on busy corners, each with an umbrella and a few flats, usually of strawberries. At the end of the day, the van returns and picks up the anonymous workers and collects the money as well as any unsold strawberries.

The waste-hating farmer takes his ripest fruit to the moving city— the road. And the herd pauses there. And grazes.

---

fruit. People would come out in their trucks and just fill the beds up with fruit. But now, no one does those things anymore. At most, someone stops by to get a peach for lunch. Our average sale now is three dollars."

In the valley, roadside stands are facing the same pressures that farmers across the country have felt for years. Get bigger, get out, or find a niche market.

THERE ARE ROUGHLY THREE TYPES of roadside stand. The first, and for our purposes, most interesting, is the on- or off-farm stand—a stand owned by a farm family, either on their property or elsewhere, that sells mostly nuts, fruits, and vegetables they grow themselves. Most of these stands are seasonal.

ON HIGHWAY 70 NORTH OF MARYSVILLE, where on clear days both mountain ranges can be seen from the Valley floor, the Lovingly stand has changed with the times. This stand too began as a way to sell excess, overripe, or imperfect produce to neighbors. As demand grew and a simple card table and tin can for money no longer sufficed, Mr. Lovingly, now passed, moved an empty chicken house to the side of the road, cleaned it up a bit, and began offering a larger variety of produce, including some from other farmers.

A newly returned daughter, Marsha, has transformed the old chicken house into a cozy storefront decorated with green cloth, old butter churns, and antique tools from the barn. The pears, apples, eggs, peaches, squash, onions, and plums are arranged with care and love into attractive piles, and a display of Dirks honey is featured.

PLATE 1.16

STEPHENS FARMHOUSE, JUST SIX MILES south of Yuba City, has evolved past the roadside stand but is still solidly rooted in that world. The Stephenses got into the roadside stand business when Ma Steph, a widow with nine children, set up a table on Wilson Road in Sutter County to sell her excess melons. The kids became involved and the stand became larger as it became more profitable. The stand eventually moved to Highway 99, and then a few years ago it took over a bakery a few miles away. Now, in addition to selling melons and other fruits and vegetables from what remains of Ma Steph's garden and from the farms her children now own, they also use the produce to make jams, jellies, cookies, buns, pies, and the best cornbread muffin I have ever tasted. We arrived when Cherie, a member of the Stephens family by marriage and owner of Stephens Farmhouse, was teaching a fourth-grade class how to make pies. Oh, the brats were everywhere, elbow deep in dough and apple filling, watching in rapt attention as Cherie taught them how to make the crust, how much apple filling to scoop in, and finally, how to seal the top. Each child carved his or her initials into the crust and then went out to the adjacent play yard to eat lunch while the pies turned hot and brown.

"It's important that kids growing up today are in touch with the countryside, that they know where their food comes from. That's why I host these field trips. So many of these kids have no idea that the peaches and walnuts they buy at the store were grown almost in their backyard," said Cherie, washing sticky proto-pie from her hands.

"The thing that makes us special, other than the pies and jams, which I do just because I hate waste and can't stand to throw away ripe fruit, is the melons. I call them 'nostalgia melons.' Ma Steph planted all these kinds of melon no one sees anymore: Crenshaw, Black Diamond, Ambrosia, Persian, Casaba, and the Pershaw melon, a hybrid we created. Old folks sometimes wander in and

"We have movie stars that come through on their way to the mountains. Everyone just loves what we've done."

The roadside stand signs, all of a kind with thick, hand-painted lettering and enthusiastic enticements, here display even more verve than usual. The Lovinglys are determined to keep their stand, despite difficult economics and the pressure from apparently unstoppable suburban development.

"There used to be even more stands around here. There was another one just down the road. It's a golf course now," Marsha said without rancor.

That this stand grew directly out of a farm is easy to see. Set behind the stand sits an old farmhouse, fenced in with chain link and harboring small, yappy dogs. A small orchard runs along the road, and an enormous red barn stands in the middle of the property, looking reproachfully at the traffic as it whirs by.

Plate 1.17

PLATE 1.18

see the melons, and their eyes light up. It takes them right back to their childhood. They can't carry all the melons they buy. Part of the reason you don't see these melons in stores is because their skin is so soft. They'll just fall apart in a truck."

I asked Cherie if any of her children were interested in farming.

"Well, Jeff, my husband, and I would never ask our children to take up farming. They should be free to do whatever they want. But...well, we have hope for our youngest, Sam. He's only seven, but he can't get enough of his toy tractor..."

We left Cherie and she went into the yard with the fourth graders.

"Now, can anyone tell me," she asked, "what grows on those trees over there?"

strawberries each when our host handed us two baskets of perfect fruit.

The man was busy, and language was mutually an issue, so we quickly thanked him for his gift and left him to his work. The Hmong stand, unlike many family farm stands, operates on a different economy. Because the small parcels of land are almost always leased, the yearly profits, after rent, seeds, and materials, are usually under ten thousand dollars. The Hmong are traditionally farmers, and their strawberry stands have changed the face of the Valley. But in a place where land prices can exceed twenty thousand dollars an acre, they have been largely unable to buy land and acquire the kind of stability that would make their influence permanent. Moreover, there is little indication that their American-educated children will have any desire to continue working with strawberries on

---

In twenty years, it is probable that the hundreds of strawberry stands scattered liberally through the Valley will have vanished completely.

MIKE MADISON HAD TOLD US that there is a Hmong strawberry stand about every ten miles throughout the Valley, and he did not exaggerate much. As Scott and I stuffed our faces in front of a taco truck in Knights Landing, we saw a hunched figure in a rolling half-acre strawberry field. The farmer gladly ceased his stoop work and explained the process. He planted two kinds of strawberries, one that ripened in April and another than ripened in November. The white plastic kept the heat and moisture in but the weeds out. The November strawberries were just beginning to come on when we were there. He demanded we take some with us. Scott and I searched diligently for ripe berries, finding some so ripe that they melted across our hands and others too green to eat. We had only collected a handful of

borrowed land. In twenty years, it is probable that the hundreds of strawberry stands scattered liberally through the Valley will have vanished completely.

WE LEFT THE BAY WELL BEFORE DAWN to meet James Saeturn and his father, Kouay, at one of their three roadside stands north of Marysville. James and Kouay are Mien, a distinct group of Southeast Asian refugees with a related language and similar history to the Hmong. Both groups are mountain-dwelling subsistence farmers with a strong and ancient peasant culture, and both fled their homelands in the aftermath of the Vietnam war.

The Saeturn stand is different from most Hmong or Mien strawberry stands in size and scope; they own two of the properties on which they farm, but they lease this roughly ten-acre plot. They sell many kinds of peppers, bitter greens, tomatoes, strange squash, garlic, onions, and of course, strawberries. Around the stand, enormous gourds and piles of expired fruit gives the structure an almost carnival atmosphere.

James took us around while his father manned the stand, stepping around two tiny Siamese kittens and smoking a homemade pipe.

"Here, taste this," James said, handing me a small, round, green berry. "These are called bitter balls. You cook them with ginger and lemongrass. They're good for diabetics." I tasted the ball and let its bitter flavor crawl around my mouth on busy centipede legs until I could take it no more and spat it out. I recalled the poem by Stephen Crane:

> In the desert
> I saw a creature, naked, bestial,
> Who, squatting upon the ground,
> Held his heart in his hands,
> And ate of it.
> I said: "Is it good, friend?"
> "It is bitter-bitter," he answered;
> "But I like it
> Because it is bitter,
> And because it is my heart."

And suddenly, my mouth parched with a bitterness of its own, I understood the lines anew. The creature who feasted on his heart did so not for the torment of it, but for the medicine. I suspect that in Crane's day, the relationship between the two was more pronounced than now. The creature, and the poet, were both looking for a cure. In my late-twentieth-century upbringing, I had never understood that medicine is bitter—not sickly sweet. And despite James's perfect English and American education, he still sees food as nourishment and medicine combined. This herb is good for digestion, this gourd for the skin. The two concepts have not been forced apart in his mind, and I was reminded of Wendell Berry's observation: what does the poor quality of hospital food say about how doctors view the human body?

James grows a few traditional Mien herbs and gourds, but many of the greens and peppers found on his land are Indian. Members of the large Sikh community around Marysville and Yuba City give him the seeds and tell him when to plant them. Throughout most of the tour, I would come across an unfamiliar plant and ask him what it was, and he'd shrug, tell me he didn't know, and then ask me to taste it. The flavors were bitter, sour, hot, nutty, and sweet.

"The Mien eat a lot of leaves. Most pea, squash, and bean leaves can be eaten, just fried up in a pan or boiled," James said, and having tasted these green parts of edible plants and found them good, I was shocked that we let it all go to waste.

"See that over there?" James asked, pointing to a tall, broad-leafed plant near the stand. "That's native tobacco. My father, he doesn't like any of the tobacco he can find here. It all tastes too harsh. Makes him cough. So he found some seeds and grew his own. Now he doesn't cough anymore."

James also hunts off his land, though not at this location. The family eats the jackrabbits and gophers that he kills, so the land also provides a small quantity of meat.

"In Thailand," James said, "they let the gophers live in the land where they grow lots of medicine. The gophers eat all the medicine and become medicine themselves. When they are fat, the gophers are caught, killed, cleaned, and rubbed in herbs. They preserve them in whiskey and sell them. These are very expensive. My father got one several years ago. It is his prize possession. Whenever an old friend visits, my father gets out the bottle, cuts off a tiny sliver of the gopher, and gives it to his friend so that he will be healthy."

The Saeturns have done well here through farming.

P<small>LATE</small> 1.19

PLATE 1.20

The intensity with which they can farm a small parcel of land rivals what the Knolls' farm is able to do, though the Saeturns, who are not organically certified though they use no chemicals, charge much less. They even sponsored a relative from Thailand to come here and work for them, learning farming and going to school at the university in Davis.

"We bought him a good piece of land in Thailand, and he's farming there now. There didn't used to be any strawberries in Thailand, but now, all the California refugees are teaching their relatives or going back themselves to grow strawberries. He'll do very well, I think," James said.

"One thing, James, before we go," Scott said. "Why is it that all the Hmong and Mien stands across the Valley look the same? They are all white, with big red strawberries on the sides. Was that a design or...?"

James laughed.

"No, no design. They are white because white is the cheapest paint, and red because strawberries are red."

I imagined white strawberry stands popping up across the lush Thai landscape like mushrooms and smiled. Perhaps they were not so ephemeral after all.

———

TOM BUNKER IS A FOURTH-GENERATION California farmer. His small orange grove, one of the most northerly in California, is situated only a half-mile from the original homestead of his great-grandfather. He lives off Bunker Road, south of Gustine. During navel orange season, roughly January to April, he leaves ten- and twenty-pound bags of oranges on the porch of his father's old toolshed. Friends and neighbors in the know come by, put their money in a can, and drive away in citrus-scented cars. He is far from any highway or well-traveled road. His customers must either possess secret knowledge of his location or be, as I was, intentionally lost, having pointed my car towards a

———

Friends and neighbors in the know come by, put their money in a can, and drive away in citrus-scented cars.

———

seeming no-place to see what I might find. The sign reads:

ORANGES →

"My great-grandfather helped dig the irrigation ditches we still pull water from. They did it back then with teams of mules and dozens of people. I can't even imagine that kind of work," Tom said.

"My father," Tom continued, "was a hard old man. Supposedly, he kept a pet rattlesnake in that shed where we sell the oranges now. I talked to his hired hands, who swore up and down that whenever they went in there, they heard the rattler and backed out. I don't know. Maybe he rigged something. Or maybe he kept a rattlesnake in there. That's a strange building, too. I didn't notice it until a few years ago, but that's actually two buildings joined together. I don't know why. We don't get rattlesnakes over here anymore. The canal they dug keeps them all in the hills."

Tom Bunker's father also didn't believe in feeding any animal that didn't work, so Tom was never able to have a horse until he was an adult. The only exception his father made was for his wife's cats, because they usually caught enough mice to survive. But one year there were too many cats, and they were starving. So Tom's father took Tom out with him into the fields at night and shot light-blinded rabbits to feed the cats.

Tom's son Joe pulled up in a big construction truck and came over. We told him we were out talking to people who were trying to hold onto an old way of life.

"Hell," said the fifth-generation farmer, "I'm working seventy hours a week trying to hold on to this place."

———

Plate 1.21

PLATE 1.22

PLATE 1.23

A STRING OF ROADSIDE STANDS can be found on Highway 135 towards Lindsay, in Tulare County (they seem to travel in packs). We stopped at a handsome, permanent, wooden sign advertising kiwis. After some knocking, Sam Hurtado appeared and asked us what we wanted.

"Well, we saw the sign for the kiwis, and we're doing a book on roadside stands…"

"Oh. That's all done now. I turned off the water to the vines years ago. I need to get someone in there to tear those vines out."

"What happened?" Scott asked.

"Well, it was always my wife's dream to have a farm. We lived in L.A. all our lives, but when we retired, we decided we'd move out to the country. I asked her what she wanted to grow. Kiwis, she said. So we put in the vines. Oh, it is

We stood in the graveyard of kiwi vines with Sam. Skeletal, twisted, they intertwined in a wreath of inevitability around us.

"You were lucky to even find me. I don't live here anymore. I moved into town after she died."

Scott and I thanked Sam for his time and staggered to the car, each of us trying desperately not to think about what would happen if our wives died before us. Would we just turn the water off?

FURTHER DOWN 135 WE FOUND the Webster fruit stand. Tulare County is citrus country, and the Websters specialize in grapefruit, but they also grow oranges and olives. Vikki, a sprightly woman edging towards eighty, runs the stand with

---

We stood in the graveyard of kiwi vines with Sam. Skeletal, twisted, they intertwined in a wreath of inevitability around us.

---

great land here. The vines just took off. The fruit was as big as ostrich eggs. We used to have a stand right by the road, and she'd sit there with one of the kids and sell them to people that stopped by. But then, a couple years ago she died. That same year, the market was flooded with cheap kiwis from overseas. There was just no point in going on. I turned off the water and watched them die," Sam said. An indescribable sadness permeated his plain speech.

"Would you mind taking us to see the vineyard, Sam?" asked Scott.

He agreed, and we walked with him slowly over the uneven ground.

"I haven't been out here in over a year. It almost makes me cry, just looking at it. This is really great land. Look, see how big the vines are?" he said softly.

an iron fist. The first time I visited the stand, I made the mistake of trying to buy just one grapefruit.

"What are you going to do with just one grapefruit?" she demanded from the window of her car.

"Well, I'm on the road. I can't eat a whole bag," I explained.

"They keep. They keep. It's only ten dollars for twenty pounds. Take them home, eat them all spring."

"Um, okay," I said, in a daze, handing her a ten-dollar bill and shouldering an enormous bag of grapefruit. I remained in a daze as she drove away.

Once I got the grapefruit home, I was forced to use it. Suddenly my table featured grapefruit ceviche with scallops and fresh jalapeño peppers. I marinated chicken in grapefruit and smoked it with fig wood. I drank grapefruit juice and sent streams of it into my eyes for breakfast. I sliced the pink flesh thin and then rolled it in sugar, using it to adorn the cream topping of a Russian napoleon cake. I sliced the grapefruit in half, scooped out the flesh, and crushed it with heirloom tomatoes, habañero peppers, cilantro, and roasted melon seeds to make a salsa which I then served in the halved grapefruit rind. Before that, I didn't even like grapefruit.

Vikki was in a hurry the next time we visited her. A worker had been injured and her husband was at the doctor's, and she and the grapefruit had to get in. We tried to buy a few grapefruit from her.

"What are you going to do with just a few grapefruit? They're only ten dollars a bag," she cajoled us from atop an ATV. So we paid her ten dollars, and she sped away into the citrus groves.

scarred by the baron of the fields behind was orphaned. But these three onions did not know that their errand would go unrun. They had obstinately rediscovered their original purpose. Hallelujah! Sunward!

She seemed insignificant in the enormity of the landscape. A scratch on the lens. We were at the Valley's end. The foothills rose to middling heights a mere half-mile away, and the jagged fingers of the Sierra reached towards a broken infinity behind them. Set strangely in the corner of this unexpectedly busy intersection was an unfurled umbrella. Under the umbrella rested a flat of strawberries, their redness, like the lipstick of women, used to catch the eyes of passersby. The girl sat slightly removed from the umbrella and the berries on a folding chair. A

With the haste of an illegal circus fleeing an incorruptible sheriff, the agribusiness-run roadside stand had fled, casting aside its thin aluminum bones.

We stopped at the abandoned husk of a corporate stand at the edge of the Capay Valley, drawn by a discarded pile of squash, melons, and onions in one corner. A day before, we had discovered a mass grave of honeydew melons. Several acres were covered with what looked like ten thousand skulls—melons whose sugar content was too low to sell. Agribusiness has no horror, it seems, of waste. Here, three onions had fallen. No casserole would keep them, no quiche caress them, in no butter would they be sautéed. With the haste of an illegal circus fleeing an incorruptible sheriff, the agribusiness-run roadside stand had fled, casting aside its thin aluminum bones. The produce that was too heavy, too ripe, too rakishly

worn, hooded sweatshirt protected her from the sun. She was reading an illustrated book of Psalms in Spanish. She may have been fourteen.

Scott, who has some Spanish, asked her if we might take her photograph, and she assented but would not tell him her name, how she came to adorn this lonely intersection, or where she came from.

I crossed the busy road to her after Scott had completed his composition to thank her myself. I asked her name. *I don't speak English,* she said to me in Spanish. I left her some salami, a wedge of cheese, and a peach. We saw her tuck the food under her chair as we drove away.

PLATE 1.24

Plate 1.25

PLATE 1.26

PLATE 1.27

THE SIGN READ WE GROW EVERYTHING WE SELL, so we pulled off Highway 132 to have a look. A triptych of half-filled bins containing stone fruit, tomatoes, and berries greeted us. An older woman saw us and came over to chitchat on one of her favorite topics: the Fishers' Fruit Stand.

"I've been coming here for years. They have the best produce here. Tomatoes, you can't get tomatoes like this anywhere. They're picking some more of my favorites for me right now, the Cherokee Purples. I live all the way in Tracy, but I drive out here every month with my girlfriends. Everything they sell, they grow, right here. Did you know that? Oh, you're doing a book? Well, you have to talk to Tom. He's the owner, though his daughter has just come back and I think she'll be taking over soon." Turning to the woman running the stand, she demanded, "Girl, I said, girl, call Tom. No, do it right now. These people are doing a *book*. They have to talk to Tom now."

I winked and shook my head.

"We can wait."

Tom did eventually appear with the woman's tomatoes, and she introduced us.

"Let me get my daughter. She's better to talk to than me," Tom said.

Tara Fisher has an advanced degree in social work and, before quitting the coast and returning to her father's farm, was a parole officer for juveniles in Santa Clara.

"I'm still learning my way around here," Tara said as she led us through the farm, "so bear with me."

We visited the Fisher farm at just the right time. The peaches and nectarines were at the peak of ripeness, and the heirloom tomatoes had just begun to fruit. The muscat grapes, however, were still a little green.

"Those vines have been here forever," Tara said. "You should see the old Italian guys come when the grapes are ripe. They buy as much as they can and make these really wonderful wines out of them. They're such characters."

We walked through a range of shades of stone fruit, talking leisurely of our mutual decision to abandon urban life and return to farming.

"There comes a time when you realize that if you don't want something to disappear, like my father's fruit stand, you have to step in and save it yourself," Tara said.

The Fishers farm one hundred acres and sell almost everything through their roadside stand. They employ a couple Latino workers and their families.

"Juan has been here for twenty years. He watched me

> "There comes a time when you realize that if you don't want something to disappear... you have to step in and save it yourself."

grow up. Their grandma lives with them, and it seems like every time I'm wondering what to make for dinner, she shows up with a giant plate of enchiladas for us. I couldn't imagine the farm without them. They're family."

After nearly an hour of pleasant conversation and bucolic meandering, we'd developed a mighty thirst, and the heavy, dripping peaches seemed like the perfect refreshment.

"May I?" I asked Tara.

"Of course."

I walked slowly up to the first tree and sized it up. No. Somehow it wouldn't do. The second tree looked more promising. I took the lowest peach in my hand and felt its heft. No. I scanned the tree, looking, as Paul Buxman had taught me, towards the highest, sun-kept branches. There. I had to jump for it, but when I touched the peach, it fell easily into my hand. The flesh was so tender that my leaping harvest had already bruised it. I had to bite into the peach and then drink from it like a chalice to keep the juice from escaping.

"Can you pick me one?" asked Scott. "You seem to have the touch."

Taken in by the flattery, I interrogated the next row of trees with a terrier's intensity. I was moving slowly around a heavily ornamented tree when I heard a rustle in the leaves. A peach fell from one of the top branches and landed in the soft grass. Thud. I leaned over the unfledged stone fruit.

"I think you want this one," I said, having to admit that nature's touch was as fair if not fairer than my own.

———————————

WE TURNED DOWN ROAD 126 in Tulare County in hot pursuit of okra. Or at least I was. Scott has no taste for the stuff. The signs led us to a seemingly deserted stand in a barren patch of land. A buzzer had been installed in the empty stand, so we rang it. Ernest Tristao appeared from his back porch.

"We're looking for some okra," I said.

"Oh, it's been a terrible year. First the rains, then it got all hot. Last year I had okra a foot long. Now it's just these little bitty things. It's happened to everyone. There's a big tomato farmer just up the road. Has probably a thousand acres. He's got no crop. I can't get my tomatoes to ripen. They stay green, then split apart before they ever get ripe. Won't have any more okra to speak of for two or three days. You're interested in the stand? Let me get Darlene. She handles most of that."

Darlene, another spry woman wearing a garland of decades, also appeared from the back porch.

"They tell me I have to keep on this oxygen tank, but I move around like a headless chicken most of the time," Darlene said.

"You move like an old lady," Ernest said, "because that's what you are. An old lady."

"Speak for yourself," Darlene shot back.

"Nobody grows okra anymore, so we do pretty good business when we have the crop. People don't like harvesting it. It's an itchy plant, they say. The leaves have some kind of poison on them. But I just pick them with my hand. It doesn't bother me," Ernest told us as we waded through small dunes of atomized soil in which he, astoundingly, grew okra, peppers, and tomatoes.

"I'll probably put most of this in pasture next year. I'm just getting too old for all this work."

"Let me get some cherry tomatoes for the boys," Darlene said, and we followed her over to the porch.

"These are my specialty. You can't get tomatoes like this anywhere. They're just like candy."

After the tomatoes, the Tristaos brought out a pile of peppers, more tomatoes, okra, and an enormous jar of pickles. Seemingly ex nihilo, their porch table issued a cornucopia of vegetables.

"It's just been such a terrible year," Ernest repeated, shaking his head.

———————————

WE WERE DRIVING IN THE FOOTHILLS on Highway 132 east of Turlock, searching, like the Magi, for another road back into the Valley, when we ran into the roadside stand at the intersection of 132 and Lake Side Road. We'd been visiting roadside stands for a year, and the thread of our travels would soon be cut. This was our last roadside stop.

The stand radiated ease and goodwill. It was built like a park picnic structure, with a high, peaked roof and no walls. A half-dozen tables contained apples, peaches, pears, and tomatoes, as well as local jams, jellies, and honey. It was the last outpost of the Valley on the road to Yosemite.

Near the back of the stand, set below the counter where you pay, sat a large box of overripe tomatoes. Having been trained, in the course of the year's exploration, to seek out this box, I had guessed where it would be.

PLATE 1.28

FARMERS • 55

PLATE 1.29

"I'll give you the whole box for three dollars," the woman said, when she noticed my gaze.

I whimpered. This is what I wanted, one of the reasons I set out on this adventure. To take a box of tomatoes on the verge of rot and stop time for them. To make salsa, or tomato sauce, or to pickle them. To foil the conspiracy of waste. What was that, I calculated, thirty cents a pound? The tomatoes sagged under their own weight. *Make of us,* they seemed to say, *something useful. Take us now, when our ripeness flirts so with rot, and make of our passion a red lamp in the wet dusk of winter.*

"I don't know how to can, and I don't have any of the stuff," I confessed.

"Well, we have canning jars right over here. I'll sell you thirty of them for ten dollars. It's easy to learn. A bunch of ladies and I get together over the weekend and have canning, pickling, and jam-making parties. Anyone can do it."

Like most lazy men, I secretly enjoy being forced into action. I bought the heavy box of split, soft tomatoes and three bags full of canning jars, resolved to do my small part for the roadside stand. I would learn to preserve the ephemeral bounty of summer.

To be an edibilist is to take some portion of the burden of the harvest upon yourself. Edibilists are not defined by what they purchase, prepare, and consume, but what they produce. Once learned, the canning arts tend, as the harvest itself does, towards excess, to more of one thing than one person or family can use before it is no good. This natural crisis can go in two directions. One is waste, the other is outward, into the homes, mouths, bellies, and traditions of other people. In a culture of deranged consumption, thrift and generosity are deeply subversive virtues. It is not enough to eat well. To truly change the way Americans eat from the land, we must feed others well.

Plate 2.1

TWO: Milk, Meat,
and Honey

Plate 2.2

The Lopez Lamb Company

When I heard Don Lopez shout, it was like a rattle under my feet or the sudden thickening of the air with angry wings. I knew I was in trouble.

"What the hell do you think you're doing?"

I turned to see the righteous tower of him making short work of the distance between us. He was something to see, that storm of a man. And I deserved his wrath. Hadn't I followed a trail of white wool shearings up to an old shed, and hadn't I ducked my head inside uninvited? I had. Now I reaped the whirlwind. Heroism was called for. I shouted apologies and explanations and began walking just as quickly towards him, even though this only made Don Lopez, a big man, bigger and delineated the aspects of his rage more clearly. My arms swept low and wide from my body, signing the umpire's exclamation: *Safe! Safe! Safe!* He pulled up just out of striking distance and we began to know each other as more than landowner and trespasser. We read each other as one dog reads another, a benign tilt of the head, shrugging of the shoulders to appear smaller—the whole body expectant, still. Eventually his anger spent itself entirely and was replaced by a nobler emotion.

"I'm sorry. I've got a sick ram over there. I was on my way to check on him when I saw you. You can come along if you want, but I've got to see him." The ram had just endured a delicate surgery and was recuperating in a small, dark pen. Don removed the wood panel from the front of the fence and shook his head. The ram looked magnificent and calm, but occasionally kicked himself distractedly in the stomach.

"That's not what I wanted to see," Don said. "You're in trouble when they start kicking at themselves like that." He replaced the panel. I asked him a little about himself and his operation. He is a teacher at a school some distance from his small ranch on Highway 160, just west of Isleton, and his daughter Andrea shows most of the goats and sheep they raise. He's also the leader of his school's chapter of Future Farmers of America. I asked him if I could come back in a couple weeks with a photographer, and he said yes.

When Scott and I arrived, Don and Andrea were preparing a select few plump and shapely lambs for a show the next day. The barn smelled of sweet animal soap and close-cropped lamb's wool. A few kids and lambs suckled on their mothers' teats in cool, small pens while slightly older animals gamboled around a large grassy pen speckled with sun and shade from the trees growing on the side of the levee. Behind the levee, the Sacramento River sent its slow, muddy burden to the San Francisco Bay. Don was allowed to irrigate his pastures with that water as partial recompense for the taxes he paid. Don introduced us to Andrea and his dogs. He keeps three Great Pyrenees for protection and two shepherds to work the herds. When Andrea was buried well and good beneath the panting bulk of the three Pyrenees, Scott asked Don if one of them had ever caught and killed a coyote.

"They don't have to. If you're a coyote and you see one of these dogs running at you, you don't stick around to find out what's going to happen."

"How's your ram?" I asked with all optimism.

"He didn't make it," Don said, "and I don't mind telling you I cried when I had him put down."

WE MOVED OUT TO ONE OF THE pastures where a herd of sheep grazed peaceably. Don was going to work his dog for us. He commanded Sally to wait near the barn and took his position on the other side of the herd. We crouched in the middle of the field so Don could direct the sheep directly into us and so Scott could snap the photo. Sheep are nervous animals but there is a hierarchy to their fears. Highest on the list was Sally, who, though she crouched beneath the grass and did not move, never left their minds. Scott and I came next, whose intentions they could not guess, and last was Don himself. Andrea and the trio of Pyrenees did not frighten the sheep but some distance was kept from them on principle. Don raised his hand, Sally stood to attention, the sheep drew closer together, and the show was on.

Humanity's domestication of dogs seems to have given them a great and unexpected boon. Throughout our visit with Don and Andrea Lopez, Sally never took her eyes away from her master. She was ready to perform the least of tasks at any moment, and would only falter in the execution of her duties from an excess of enthusiasm. A working dog has been given Purpose and has responded admirably to that gift. When being used, a dog is overcome with an ecstatic joy unavailable to its untrained or wild brethren.

The sheep were doing their best to manage their fears. Sally, who darted forward, back, or crouched invisible beneath the grass as Don commanded, had driven them, in three attempts, quite close to our position. Don, however, had expected better.

"The man I bought her from," he said, "told me that she'd run perfect except when you try to show her off."

Suddenly the zigzagging river of sheep overwhelmed us and we were inundated by the woolly current.

"I got it!" Scott shouted. Don called Sally off. We stood up. The sheep scattered. It was beautiful.

DON ONLY HAD TWO QUADS, which meant that I, the smaller man, would have to perch uncomfortably on the back of the machine while Scott resented having to care if I fell off. It was an annoying arrangement, but what is uncomfortable to me would be cruel and unusual to Scott, so there was nothing for it. We were headed to the far end of Don's land to visit some of the goats and complete our tour.

"Do you eat any of the goats that aren't show quality?" Scott asked.

"I don't like to eat them. They're always worth more when I sell them. Sometimes people will stop by and I'll sell them one. And sometimes I go down to Escalon and sell them at the market there. Have you guys been to Escalon?"

"No."

"Oh," Don chuckled, "you have to visit that place. On Fridays they have a second auction outside and sell all kinds of things. It's something to see." We promised to check it out. Scott and Don started up the quads, I held on tight, and we were on our way.

PLATE 2.3

PLATE 2.4

The goats were friendlier than the sheep, and more individualistic, as the old Christian saying warned, but I couldn't see how they suffered for it, even if each one did vanish daily to have his or her beard combed in hell. In the farthest corner of his property, Don stopped the quad and stared back at the green expanse of his fields, the dark clusters of his goats, the white blossoms of his sheep, the red pyramid of his old barn, and the low rectangle of his own home.

"This is my favorite place. I come out here once a day and look around for a while," Don said, and then fell silent. We watched with him. One of the Pyrenees loped threateningly along a fenceline on the other side of the fields. Some minutes passed. When he was done, Don didn't say anything but simply started up his quad and looked back at us. Time to head in.

WE STEPPED INSIDE HIS HOUSE and saw a pile of trophies and ribbons set atop a woodburning stove. Scott asked if he could take a picture, but Don hesitated.

"I wouldn't want people to think I was showing off," he said.

"I'm sure anyone that knows you would know better," Scott answered.

THE LOPEZ OPERATION was as much Andrea's as her father's. She was everything I hadn't been as a farm kid: obedient, enthusiastic, knowledgeable, and patient, which is not to say that she didn't talk back to her father when he deserved it. She seemed to know the names of all the animals from fifty feet away, no mean feat when dealing with sheep, and could prepare a lamb for show or sale perfectly on her own. It actually seemed possible that she might continue raising goats and sheep as she grew older—

that, through her, this land might continue dreaming this particular dream.

"Do you think Andrea will keep this place going?" I asked.

Don shook his head.

"You can't think about land like that anymore. Andrea loves it here, and she loves animals, but she wants to be a vet. I've known a lot of girls her age that say that, but very few ever get there. She's better suited for it than most, but you never know. Even if she does become a vet, where would she work? There's not enough business to keep most large-animal vets

Suddenly the zigzagging river of sheep overwhelmed us and we were inundated by the woolly current.

working out here anymore, and to be a small-animal vet you need to practice in town. Which would mean commuting from here to there. No. I can't ask her to even keep this place. As much as I love it, I know it's just an investment. Maybe she'll keep it, maybe she'll sell it. That will be her choice."

Don had certainly made peace with the idea, and any uneasiness I sensed as he spoke may well have been my own. I was uneasy, and I am. The language of these choices is one of freedom, but how free are they, or have they been, the children of ranchers and farmers? If you choose to leave a place because it will probably be taken from you anyway, because you cannot make a living there, because you have to leave for college, because everyone you know has left, do you do so freely? No. You were coerced. The mainstream disdain for rural life no doubt contributed, but it is really because of the economics of the past hundred years or so that farmers and ranchers must consent to their own decline. Or to quote Kris Kristofferson: *Freedom's just another word for nothin' left to lose.*

Plate 2.5

Save the Star Thistle!

We arrived at Debbe's with less than an hour left of the day. Scott's habitual expression of watchful calm had vanished, replaced with a look that was agitated and predatory. Sunlight, in its slantwise leave-taking of our bit of earth, is honey, meat, and milk to photography, and photography is always hungry.

Ducks and geese patrolled Debbe's front lawn; two goats and a frolicsome kid lifted their eyes from the grass to my car. We ran to Debbe's door, not knowing that we had found the Genuine Article.

On a two-acre plot outside of Knightsen, Debbe makes honey, pours it into bottles and tubs, and lovingly arranges those bottles and tubs on a card table on her front porch, allowing people to trade honey for money as they will. She has also begun experimenting with goat cheese, which, prohibited by law from selling, she gives to neighbors, one of whom had led us here. We studied the simple economy of the honey's arrangement on the table while listening to the muffled sounds of someone busy finding her way to the door.

When Debbe answered, we rushed through our standard bit of vaudeville about the book, and she let us inside with a friendly caveat.

"Just don't take any pictures inside the house. They're liable to shut me down."

We are received with enormous and reflexive generosity throughout the Valley, as if the openness of the landscape grafts itself onto its inhabitants, and Debbe was no exception. This generosity pleasantly bewilders Scott but haunts me. I sense something ancient in instinctive hospitality, a worldview that predates and outshines civilization. When I am the recipient of these gifts of time, food, and character, though sincerely grateful, I am inwardly pensive. There is in the instinct for hospitality, most commonly encountered among the rural poor of any country, the seed of a human destiny unfulfilled. When expressed, its usual absence sounds itself all the more, like the howling of the pet wolf who knows nothing of the pack.

Debbe's home suffered from an only average dishevelment. The house had recently been moved to this lot and evinced the unavoidable decrepitude inherent in the shoddy building materials of thirty years ago, but we didn't stay there long. To be inside with this light? Never! Debbe was in a hurry as well. The milch goat was beeeing for her evening relief, and there were other chores to do. Debbe whistled in the goat while Scott installed himself in the corner of her outhouse-sized milking shed with his camera, and I held the gate open. The other full-grown goat grazed sleepily while a kid suckled at one of her teats. When the kid was bigger, Debbe said, she'd butcher them both. The milch goat trotted onto the platform and tapped a hoof with restrained impatience.

Debbe sat sidesaddle on the milking platform—she needed no stool—and set to work. Her hands squeezed and tugged with a rhythm deft and sure as she half-filled a stainless steel bowl with a soft white liquid that frothed and radiated warmth. Her goat stood at attention with the carriage of one for whom duty is pleasure, and pleasure duty. When she'd been milked dry, she trotted back to the others and gave a happy little kick. In but two days, her milk would be fresh cheese.

The only pressing chore done, Debbe relaxed and began to tell us about her place, making gestures that

eight-hundred-pound steer, and from the ardor of her description we could tell that their visit was one of the holidays of her year.

Debbe's relationship to her animals was guiltless. She talked with perfect equanimity of skinning, preparing, and cooking animals she had lovingly fed for a year or more, using the kind of nonchalant tone usually reserved for discussions of the weather. This shocked me. Of all the places Scott and I have visited, Debbe's little bit of land recalled most the Kansas farm where I grew up. In many of her eccentricities, she reminded me of my mother, but here she deviated. We too kept, and still keep, many animals meant to be meat—but ours never make it to the butcher. My mother will go without sleep for days nursing a baby bird or a hurt mouse. If forced at gunpoint to butcher

She talked with perfect equanimity of skinning, preparing, and cooking animals she had lovingly fed for a year or more.

swirled the fresh milk wildly about but never spilled it. A beautiful Black Angus steer, Olympian in every other aspect, stared at me with pure terror, as if I was the specter of his upcoming slaughterer. As I was to him, he was to me: a specter of the upcoming slaughtered. Before our travels were over, Scott and I were determined to witness the death of an animal meant to be eaten. I looked at the steer, and he at me, and we shuddered. His mother, who stood next to him in complete, cud-chewing calm, was serviced every year by a local bull, and every year her offspring provided Debbe with all the meat she needed and more. Debbe butchered every other animal herself—rabbits, turkeys, goats, ducks, and geese—but she would rely on two brothers, roving butchers from Modesto who dressed and walked and talked just like cowboys, to handle the

and eat one of our animals, mother would unblinkingly take the bullet. She is not a vegetarian, and has not much examined the morality of eating. I think her feeling for animals is a tribal one. Any animal she is able to help, or has known, somehow belongs to her and is exempt from the knife. All other animals exist outside of her emotional hunting grounds and can be eaten without qualm.

My own thoughts are more complex. I have inherited my mother's disease of compassion, but am not reconciled to my omnivorous ways. I have had two short bouts of vegetarianism and both ended not by moral argument but by the weakness of the flesh. Once, an herb-crusted Cornish hen at a familiar, friend-filled table, and another time, after two days of spleen, faithlessness, and bile, I broke down with two bottles of wine and a rare

PLATE 2.6

PLATE 2.7

twenty-ounce porterhouse slathered in horseradish. I am too much a Karamazov to renounce the various pleasures of eating and preparing meat, and I have learned that my grand passions cannot be repressed, but only harnessed to a fiacre and driven wildly about. And yet I require a moral justification. Eating meat is a humbling pleasure to me. It binds me to a world my reason has been trained to despise and seek freedom from. To be nourished by an animal is to admit my limitations, my place of dependency in nature. Creation, in all its animal mystery, and the necessity it places on me, are nothing to forsake, but a rare opportunity to find peace from an almost universal vanity. I did not learn to eat through my enemies, but through my mother, grandfather, and the kindness of friends and neighbors. Nourishment is, in a sense, family, and like my relationship to my family, my relationship to meat is difficult, impulsive, and an excuse for growth.

The vegetarians are still the better people, and the meat industry a monument to human cruelty, but the omnivorous passion, properly expressed, can also serve moral ends.

Debbe, in this sense, was blameless. Her animals lived happily and loved her as well as any animal may, and when she needed them, she killed them quickly and ate them with pleasure. I realized that it was her innocence that disconcerted me most, as it is something I feel I can never attain.

The animals were all fed and the sunlight was nearly gone, and Scott was anxious to see the honey fields. Debbe walked with a step that grew ever lighter as she drew nearer the hives. The rest of the farm, her vegetable garden, the milk, eggs, and meat provided her with sustenance, but her honey did much more. The bees offered her luxury, money, and interaction with the outside world. It was February, so the bees still slept, but as she stared at the piles of white boxes, we could tell she was seeing them at the height of summer, and as she described their inhabitants, legs gilded with pollen, floating heavily around the hives, their wings humming with the noise of a thousand flamenco guitars, we could see them too.

"What kind of honey do they make? Where do they go?" I asked, having developed, at least in my own mind, the ability to discern wildflower honey from clover, orange blossom from alfalfa.

"Oh, who knows where they go? Everywhere, I guess. But they come back and make the best honey in the world." The words "Best Honey in the World" were, in fact, printed on her labels. The sun went down as we talked, and Debbe invited us back in the house. With the honey she doesn't sell, she makes mead, and from the mead she doesn't drink, vinegar. She uses most of the goat milk to make chèvre frais (which translates less poetically into "fresh goat cheese") and what she calls "moldy cheese," which has a hard, veined, white rind and has been aged.

Her moldy cheese was gathering the weeks in a closed porch behind her house that also served as a workshop. The handle of her hammer had grown a knot of wax halfway towards its head, where a small woman would most comfortably grasp it after working with her hives. She cut us each a slice of the aged cheese. Its sharp, dry, goaty flavor made us thirsty, and we moved inside.

Mead was fermenting in several twenty-gallon bottles underneath her kitchen table, each bottle fitted with an intricate bit of tubing fitted into the top that relieved the pressure. She counseled us to hush.

"They talk to me. Shhh and you can hear them."

A few long moments of expectant silence passed, and then the mead found its voice.

"Gung-gu-glunk?" it inquired, the sound echoing bright and round off the sides of the plastic bottle. Debbe was giddy as a new mother at her mead's performance. In order to find "the good one," she had to tip each bottle over and pour out a taste, often swirling with bits of wax. Scott and I each had a canning jar from which to drink, and sampled most of the bottles along with Debbe. Each batch shared an alcoholic strength, but the flavor, thickness, and shade of amber varied greatly. The alcohol went immediately to our heads, clandestinely preparing us for a transformational experience. Finally, she came upon the object of her search. As soon as the liquid conformed to the shape of our jars, we could tell that this mead wore a different cloth than its brothers—that it was a drink as fair as Joseph. The liquid shined like polished mahogany. A delicate froth burst audibly on the surface, releasing a rich, yeasty scent. Scott and I raised the jars to our lips with reverence and sipped. An autumnal flavor emerged, as if the honey we had always known and loved was but a child—this was a taste still unquestionably honey, but aged and hoary, a honey with wisdom and forbearance, delicate as an old man's bones or the powder on a moth's wings. Debbe, oblivious to us, gazed at her creation with an admiration softened by the fondness that often accompanies friends of long standing.

"It's in its second fermentation," she informed us, and then moved some papers around on her table so that we would have a place to sit down.

As the mead warmed us, our questions became languid and her storytelling more intimate. We praised the tidy abundance of her life and suggested that it must be a dream come true. She surprised us and shook her head.

"What I wanted to do was live in the state parks in a tent, gather roots and berries, and hunt the wild boar." Her parents, frightened by this plan, recalled the empty lot they had inherited. Wouldn't she rather, they asked, live off her own land? They moved an old prefabricated house onto the lot, and Debbe moved in with reluctance. That was five years earlier. Her family also provided her with a small

sum of money every month, but Debbe's expenses were few. She had become interested in film and was taking an evening class. She'd just watched *Rashomon*, and Scott and I made casual conversation about Kurosawa, Bergman, and Tarkovsky.

Debbe showed us how she made popcorn from last year's maize, rubbing two dried ears together until the hard kernels fell into a metal pan with a sound like hail stones on a tin roof. She poured several kinds of mead vinegar from small, narrow, corked bottles. This, she said, was her favorite. A heroic contest between sour and sweet played out in our mouths and both, miraculously, came out the victor. Debbe gave us some samples of her fresh cheese and when I asked how she learned to make it, she handed me a book, aptly titled *How to Make Cheese*, by a woman in Canada. Debbe had found some confusing information in the book and wrote the author a letter full of questions, to which the author helpfully responded. Debbe opened her cabinet of home-canned vegetables and sauces, eaten down to a quarter of their original number as winter drew to a close. We were getting the full tour from a woman almost completely self-sufficient in her food production. She even made her own soap.

A metal pan full of yesterday's milk sat behind us on an unlit woodburning stove.

"I think it's about time," Debbe said. "Do you want to see me make some cheese?"

As she set up her equipment—six small plastic molds with sievelike holes on the bottom, a metal pan with a grate over it on which to set the molds and allow the whey to drip into the pan, a slotted spoon to scoop the cheese from the pan into the molds, and a light to see by—she rhapsodized about the sound of the process. The tinkling of the whey as it drained into the pan below reminded her of rainstorms and wind chimes. She enjoyed this music more than even eating the cheese. Her preparations were complete. I watched silently while Scott's camera restlessly measured the composition and light. The whey symphony began quietly, the few whey-drops from one mold making a small and uneven series of sounds, but it was soon joined by an increasing multitude, and the room became filled like a glass with the wet tintinnabulation of gentle music. Forgetful of our presence again, Debbe giggled and grinned, completely giddy. As the drops of whey began to slow in frequency, one of the bottles of mead added "Hung-na-flump?"

I asked Debbe if she used the whey and she looked startled, like I'd just discovered a secret.

"I use it in everything," she confessed. "I've begun to make my tea with it, and I drink it straight." I gave her a Slavic recipe that uses whey as the base of a cold soup, and she vowed to try it soon.

I bought a comb of honey and she gave us a block of fresh cheese. As we stood to leave, she gave vent to one of her annoyances. The rant contained the same pragmatic impracticality that characterized the rest of her life.

"Everyone hates the star thistle. But I know it makes the best honey. My neighbor has horses, and she's always complaining about the star thistle ruining her pasture. Horses can't eat it. But goats can eat it. Bees make delicious honey from its flowers. I want to get bumper stickers that say 'Save the Star Thistle!' What do we need horses for in this day and age, anyway? You can't eat them."

The mead had worn off and left Scott and me with sharp little headaches, vocally empty stomachs, and almost an hour to Los Banos, where we hoped to find a simple repast and repose in some seedy, highway-side motel.

With only forty minutes of daylight left in what had already been a meaningful day, Scott and I had stumbled, unwittingly, upon the Genuine Article. Edibilism is expressed throughout the Valley in myriad, but always partial, forms. Debbe had created a life in which she was intimately involved with every aspect of everything she ate. Moreover, this lifestyle had evolved well past simple survival. It had a

culture and aesthetic all its own. Her food was filled with beauty and rare flavors. Her cheese, mead, and vinegar, if she was able to sell them, would command gourmet prices. But who was Debbe? An eccentric urban refugee, without heir, needing no employment, who had begrudgingly chosen this life instead of living in total wildness in the state parks, hunting the wild boar. Her presence in the Valley was unique and impermanent. Her lifestyle was very distant from contemporary values, and yet, still completely viable.

> Debbe had created a life in which she was intimately involved with every aspect of everything she ate.

Imagine if more people produced their own meat, cheese, vegetables, and even alcohol. Imagine three hundred not just lone individuals, but families, with this way of life lived in the same town. And now imagine one of these towns replacing every Willow-Tit-Willow-Tit-Willow development scheme across the Valley. Debbe expresses the true human potential of the Valley, of all rural life, but in a guise so unfamiliar as to be nearly unrepeatable. What abominable education have we received to cripple our ability to create a natural and ancient community of insects, mammals, fowl, and plant life, all conspiring to supply us with a life of meaning, pleasure, and autonomy? Why is this road only traveled by outcasts, loners, and oddballs?

The road to Los Banos was lonely and dark. After reliving our favorite moments of the day, we retreated into the companionable silence inherent in each of our natures. As the lights of Los Banos became visible, Scott spoke.

"In the morning, we need to get some crackers for that cheese."

PLATE 2.8

Like Sharks on Bloody Meat

To a tree branch clung a suit of heaving chain mail, of movable type, one creature cleaved to another creature indistinguishable from itself, and it to a hundred more, their combined activity forming a dark pustule of forty thousand wings, two hundred and forty thousand legs, and twenty thousand stingers; the bees invaded the yard of young Harold Dirks in Marysville, California. Harold watched the swarm with intent and listened to its drone. Gravity summoned the colony syrup-like towards the earth, only to be thwarted by the communist energy and intelligence of many winged tailors constantly gathering up the hive. Harold's father, a carpenter, built a large box, removed the bee-laden branch, then placed the errant colony into the box and covered it with a lid. Young Harold made a pilgrimage to the public library and walked away with all the books they had on the apian arts. He bought his first hives and combs from a local feed store but built subsequent hives using the tools and scrap from his father's woodshop. The origin of the bees was a mystery—but to the supra-rationally inclined, the hand of fate pinching the colony between its fingers and hanging it on the tree like a dark lamp is not difficult to see. Harold swallowed destiny's hook. A beekeeper was born.

Decades have passed since that mythic encounter, and Harold is more enthralled by and entrenched in the bee business than ever. We met him at six-thirty in the morning at the Eagle's Nest in Marysville, and after the important business of ordering breakfast was dispatched ("Steak and eggs," "Steak and eggs," "Well, I guess I'll have the steak and eggs"), we asked him how the year, as of the end of August, had been.

"Hell," said Harold, "I haven't had a day off since Easter."

The trouble was these damn mites, a new parasite that attached itself to the bees and sucked the life out of them, coming, according to the word round the honeycomb, from Asia. In addition to his duties as a bee-herd, Harold worked full-time as an inspector for the California Department of Agriculture. He tended his hives every day before and after work, and then all day on the weekends. In terms of busyness, the bees had nothing on Harold.

"We're going to take a vacation soon, though. Go up through Oregon. I owe it to my wife," he said, looking with weary longing at the motorcycles driving past us, heading north on Highway 70.

HAROLD LIVES WITH HIS SECOND WIFE amidst thick kiwi vineyards outside Marysville. His yard is laid out like a working yard—train, stock, ship, or switch—with two rows of surplus USPS Air containers holding excess hive materials and blocks of unused beeswax; rows of oil drums on slats, each row storing a different kind of honey—wildflower, clover, star thistle, sage, the barrels protected from the sun by slats of corrugated tin; and a forklift to move it all around. The honey is bottled in an oversize garage. The labeling is done every couple weeks by his daughter, who travels up for the chore.

technique of a barman, and for the same reason—a slow pour keeps the air bubbles out of the barrel and prevents froth—when Scott asked him a question.

"Have you ever just walked away from one of these buckets and let the honey just overflow and get all over everything?"

Harold looked surprised.

"Have I ever...shit. I'll tell you what I have done. You know that big barrel out by the centrifuge? I've let that thing go, and lost the whole thing. That's about four hours' work with a snow shovel to get all that honey cleaned up. Have I ever...I've done that twice." He chuckled to himself for a

---

Staring for nearly a minute, I finally began to see tiny black specks hovering in the air, and then I saw more.

Harold's primary customer is the network of roadside stands that extends north to Chico and south to Sacramento. Harold takes honey to every stand once or twice a week, depending on the needs of each. He also supplies a natural grocery in town that sells his honey in bulk, and a local Greek woman who opened a restaurant after her husband left her. The reach of Harold's small empire is surprising. The desires of his customers are so varied that if he restricted himself to just the honey from his own hives, which mainly collect nectar from wildflowers, star thistles, and safflowers, he would go bust. So he trades wildflower honey for sage honey, which can only be collected farther south. He purchases outright the much-prized Montana clover honey. He has also just begun to purchase honey sticks from an outfit in New York. Hawking a honey stick for a quarter pays off. The heartbreak of good honey must be indulged.

Harold was pouring a broad stream of star-thistle honey down the side of a white five-gallon bucket with the patient

minute or two and then became grave. "But that's not the worst thing I've ever done. I used to have this Jeep that I'd use to transport the hives and combs. One night I must have been really tired, and I left all the combs in there. Next day they were in the sun. With the windows down. All day. I got home and saw half the house covered in bees, the sky dark with bees, and then I thought, 'You dumb shit' and ran over to the Jeep. Bees from all over the county had swarmed onto that honey. It was like sharks on bloody meat. I just dove into the mess, rolled up the windows, and then ran inside."

We'd finally killed enough of the morning to go out to the bees. The sun has to warm the wax seals enough for Harold to be able to pry them off one another. He was excited to try out a new method of opening the hives, a piano wire with a wooden handle on each end.

The honey yard sat along the Feather River, just outside of town. Dandified attired joggers from the new housing development nearby ran back and forth along the dike with young dogs and strollers. Harold parked his flatbed pickup a short distance away from the hives. This location suffered from two scourges, bored adolescents who would come by and knock over his hives just for the banal thrill of it, and skunks. Harold didn't explain right away what the skunks did to the hives, but I thought I knew. After all, what is a skunk but a tiny bear with a chemical warhead strapped to its ass? The skunks were after the honey, like the bears, like us.

The skunk smell hit us as soon as we tumbled out of Harold's rather cozy truck, and the skunk itself was spotted soon after, roughly two weeks dead, near Harold's front tire. He checked the trap (empty). When it's full, he shoots the skunk painlessly in the head and leaves the corpse there to deter both other skunks and teenagers.

"I used to just tie a rope to the trap and throw it, skunk and all, into the river to drown it and keep the smell away. It maybe wasn't the most humane way to do it, but that's what I did. But once I went over to the bank and gave the trap a good heave, and just as it was arcing towards the water, I noticed two fishing boats, one on either side, looking at me curiously. It didn't take long for the drowned skunk to float up and start drifting towards one of the boats, bringing that stink with it. Oh, they started shouting...Anyway, I just shoot the skunks now."

"What did you do," Scott asked, "when the boats saw what was going on?"

Harold chuckled.

"Oh, I ran. I just got the hell out of there and didn't look back."

Harold brought gear for both of us, white, full-body overalls and the standard netted hats, and while we were suiting up he got his smoker going with rags soaked in engine drippings and hay, casting a practiced eye on the hives from a good fifty feet away.

"Good activity. I guess the new mite treatment is still working," he said with pleasure. I cast my eyes in the same direction and saw a dead and lifeless grid of maybe a half dozen white hives. Staring for nearly a minute, I finally began to see tiny black specks hovering in the air, and then I saw more. Harold had taken them in with just a glance.

Harold popped open the top of the first hive using his garrote, happy with the expediency of this innovation. The hive thrived and Harold extracted comb after comb of golden honey and held them up for us to admire. The bees had been lightly smoked, but it hardly seemed necessary. The one rule was not to stand in front their entrance; this would create a traffic jam and the bees would react with wing-rage, much like human motorists.

"Look at this," Harold said, pointing to thin scratch marks in the dirt beneath one of the hives. "I need to move the trap." The hive seemed undisturbed, despite the skunk's recent visit, and my "skunks love honey" theory became questionable.

"What..." I asked with hesitation, "do the skunks do to the hives?"

"They come at night and make a bunch of noise to disturb the bees, scratching the ground and smacking the hives. Then, when the bees come out to see what's going on, the skunks eat them, just snap them right up."

"They don't care about the honey?"

"Nope, just the bees."

Sonsabitches, I thought. *Sneaky, stinky, crazy sonsabitches.* My respect for the noble skunk, already high, went shopping for a stepladder.

Harold carefully inspected each of his hives, sometimes shifting a thriving cell to a hive that needed support, and once reintroducing a queen to a sterile colony floundering in democracy. Some of the unused boxes were brought back to the truck to be stored, a few bees stranded on each. It was hot as hell, and as the bees were feeling fat and happy that day, Harold said we could shed the suits, provided we kept the masks. We did so with grunts and loud sighs.

We were just about to leave when Harold gave us a short, illustrated lecture on how bees make honey.

"The bees have what's called a honey gut. That's where they store the nectar they get from flowers, and it causes an enzymatic change, turning it into honey." While saying this, Harold casually plucked one of the now ubiquitous bees out of the air with one hand and ripped it partially in half. Suspended between both halves of the bee hung a small jewel of honey. "That's the honey gut. That's pure honey in there." And then Harold grabbed the gut with one finger, discarded the rest of the bee, and scooped it into his mouth. "Mmm."

Scott was changing film and didn't quite see the extraordinary event clearly.

"Did Harold just catch a bee out of the air?" he asked, hurrying over to where we stood.

"Yes, and pulled it apart to show us the honey gut."

"Can you do that again, Harold?"

"Sure." And with equal ease, he snatched another bee from its innocent flight and, as if he was untying a knot, pulled both ends of the bee away from one another until the honey gut hung full and fat as a tick between them, catching the sun, burning with light. This one found its way into Scott's mouth.

"Mmm. That's honey," he said.

What is the life of a bee worth? I wondered. The management of a hive and the health of the queen—the only necessary personality—depend on sheer numbers. One bee, or two bees, or three bees, are of no value. The hive requires hundreds of workers and drones to ensure that enough honey is stored to last the long, cold winter, and that the queen is nourished long enough to produce an heir. Every action taken by a beekeeper is communal, save for adding a queen to an abandoned or infertile colony. Centuries away and thousands of miles distant from any monarchy, I was suddenly swept up into a Homeric drama, where a hundred thousand nameless souls could and would perish in defense or attack of Helen's bedroom whims. And who was this standing next to me—Harold no longer, but Zeus, looming over each buzzing city-state with total, if benevolent, power. The singular defense of the bee is an instrument of its undoing. She may sting once, then die. The beehive, a civilization of dizzy grandeur, an archive of eternal spring, is constructed piecemeal of endless tragedy. *Here they are,* I thought, watching them break through the invisible trade-lines between many hives and the barbaric, floral beyond. *Here are the Greeks.*

We departed the first cluster of hives and headed out for a set Harold had leased to a cucumber and squash farmer. On one side of the field sprawled the ruin of a prune orchard. This year the fruit came on so thick and strong that some of the trees had shattered from the weight. Still more had lost entire branches, and nearly all of the trees were propped up with two-by-fours. Across from the prune orchard a brand new mosque raised golden onion domes and a single crescent moon into the Valley sky. I picked soft, dusky prunes off the ground while Harold pointed out invisible bees to Scott, who then chased them with his camera.

"There's one."

Shuffle, shuffle, crouch, pause, focus, curse.

"There's one."

Shuffle, pause, step, crouch, focus, curse.

"Over here."

Eventually, Harold grew impatient with the laziness of these bees. We'd eat lunch at Opa!, the Greek place that uses his honey, then head out to the other fields until we found Scott some bees on the wing.

Despina's husband had just left her, and Harold shook his head.

PLATE 2.9

Plate 2.10

"I just don't know how a man could leave a woman that cooks that good."

Opa! resembled just another strip-mall fly-by-night, lit by long tubes full of glowing chemicals with a soft-drink dispenser in the corner, but Despina quickly put proof to the lie, and we were soon following more orders than we were giving. She was alone today, her daughter being sick, and our food took some time to arrive, but when it did, the waiting dissipated from our minds like spit on a Coleman stove. I don't know much about real Greek cuisine, but I know a bit about cheap, shopping-plaza Greek food, and this was the best I'd had. It was suspiciously good. I asked Despina where she sourced her ingredients. It was the truck, right? These days it is always the truck: one of the nation-spanning network of big trucks bringing canned tomatoes and bags of ground, frozen sausage to almost every restaurant on every street in every town, suburb, or city in the United States.

"I go to the swap meet every weekend and buy most of my fruits and vegetables. That's where you find the freshest stuff. And then there is this crazy Greek guy who has some kind of a garden. A couple times a week he leaves me a big box of beautiful produce. I've only ever met him twice." Aha! Was an edibilist act behind every tasty thing in this valley?

We finished our meal with baklava and puffballs, each tied up in a sack and tossed from a bridge into a river of Harold's dark mercurial honey. On our way out, Despina gave Harold a five-gallon bucket that once held feta cheese to refill with his product. Harold was excited by her thrift.

After lunch we drove deep into the field lands of Sutter County, a strange, though—in the Valley—common landscape far from any well-traveled roads, sporting no homes, just a patchwork of melon, squash, tomato, sunflower, and safflower fields and prune and almond orchards, each separated by an unending series of irrigation ditches and connected by ridgebacked, potholed dirt byways. The first two fields were a miss, but the third was not far away. Having, for some time now, been snuggled between Harold and Scott in Harold's thankfully automatic pickup as we bounced along the dusty, abused roads, I conceived a fierce desire for the relative freedom of the bed of Harold's truck. I convinced him I wouldn't throw myself under the tires, hopped onto the flatbed, and found myself a secure place to crouch near the empty combs we'd taken from the first site, reveling in the fresh air and the noise of the outdoors as we bounced along to the safflower field.

The beehive, a civilization of dizzy grandeur, an archive of eternal spring, is constructed piecemeal of endless tragedy.

My hand exploded in pain and I looked down to see a strange, spent-looking bee waddling slowly away from my knuckles. The end of her thorax was torn off. *Stupid bee*, I thought with sympathy. She must have been clinging to the empty box for hours, hopelessly far from her fellows, lost to her queen, angry, confused, and disoriented. Did this sting give her short, throwaway life meaning? By choosing the time and place of her definitive act—by doing the one thing that could distinguish her from another bee—did she take a brief, final moment of control over her otherwise impersonal life? And, more importantly, was I allergic to bee stings?

Hornets, mud daubers, yellow jackets, bumblebees, fire ants, and army ants had all injected me with their venom, and none had caused any more than the usual amount of pain and irritation, but the honeybee had never martyred herself on my flesh before, and I was curious to see how far the swelling would go. Moreover, hadn't Harold just told us a story about an old beekeeper whose bee allergy only kicked in after fifty years of enduring stings? These things could set in at any time. I could be struggling for breath in mere minutes. Fascinating!

A safflower field is a beautiful, Martian thing. Warm, generous red and orange hues shine forth from miserly blossoms stuck atop thick stalks with needlelike leaves. The field spread out across dozens of acres before us, the independent shuddering of each flower making an arrhythmic discord. Harold let out a sigh of relief.

"Here they are. Look at all of them." After a few seconds of deep glaring, I saw a couple bees bouncing from safflower to safflower. "There's hundreds of them here."

Scott and Harold resumed their interrupted chase and I examined the progress of my sting. There were no striking, alarming reactions, just a steady painful throb and a new difficulty moving my joints. It looked like I wouldn't need a trip to the emergency room after all.

Scott and Harold's work ended victoriously. We all clamored back into the truck and pointed it towards home. Evening approached; our day with Harold was drawing to a close. As we drove past the kiwi vineyards and pulled into his drive, Harold directed us to a row of trees behind his house.

"See those? Those are freestone peaches. The best peaches in the world. You can't hardly find them anymore. But only the best for us. Let's see if Mom's got a treat for us waiting inside." Scott and I, groggy from the heat, the drive, and a day that had begun at four in the morning, suddenly perked up and exchanged an intense look that shouted: *cobbler.*

Inside, our noses confirmed our suspicion. Over bowls filled with warm cobbler and topped with ice cream drizzled with Harold's wildflower honey, we discussed some of the other uses of honey.

"I've heard that local honey is good for allergies, and people ask me about that a lot, but I tell them that I don't give out medical advice. But honey's just like WD-40; you can use it for everything. My wife heard it was good for the hair, too, and once she took some into the shower with her. She got the worst tangle. The next day a guy I work with told me his wife tried the same thing!"

"What's it like," I asked, "being the wife of a bee-keeper?"

She gave Harold a long, warm look. The love between them was palpable, thick, sticky, and sweet.

"Well, I don't see him that often. But I know he needs those bees. It makes him happy. And I don't mind the honey, that's for sure."

"Besides," Harold said, with a strange light in his eyes, "We'll be on vacation soon. I can't wait. We'll follow the Columbia River all the way to the coast and go through all those orchards there. I want to do what you guys are doing. I want to see how other folks live for a change."

Harold gave us some honey and fresh peaches for the road and we headed down to Davis for our evening date with some farmers' market fruit and vegetable pops. Scott noisily devoured one of the fresh freestones.

"This is not the best peach I've ever had, but it's a minor player in the same league," Scott quipped between wet mouthfuls.

Harold had envied us our travels. To be envied by a man you yourself envy, and to envy even his envy, is a rare and pleasant event to contemplate, heading south on Highway 113. Even that which seeks to divide us binds us. We are bound. As the bees are bound, we are bound.

Bubbas

Friday is small animal day at the Escalon Market on Lone Tree Road in San Joaquin County. We sought it out on the advice of Don Lopez. The auction had only started when we arrived. The parking lot was full of shiny pickups and their dusty companion trailers parked at acute angles in relation to the auction house. I slid the Jetta into an empty spot and we strolled up to the sale barn, me with my notebook and Scott with his Rolleiflex and tripod, as if we owned the place.

I would occasionally accompany my mother to the auction house in Salina, Kansas, when she needed to buy a new saddle or sell an old one. I enjoyed attempting to crack the impenetrable code of the auctioneer, drinking coffee with the old cowboys in the café, and the inevitable moment when Mom would be compelled to rescue a hurt-foot colt, an old goat, or a pair of runt piglets from whatever ignoble demise awaited them after sale. My favorite of these rescues was a she-piglet who I christened, in an early display of poetic acumen, Pig, and who finally died at an arthritic old age after I had left college and come to California. Would I be able to resist following in my mother's footsteps, even considering the complete impossibility of maintaining a barnyard animal of any size in my Bay Area apartment? And what would my wife say? The act of lunatic compassion would be worth it just for the look on Olga's face. A goat, perhaps. Just a little one.

The architecture of an auction house varies little, and pleases as those structures of pure utility so often do. A dirt-floored pen with automatic doors on either side sits at the base of steep, concrete amphitheater seating taken up by small groups with much space in between them. The auctioneer is installed well above the pen, facing but some distance away from the bidders. The entrances for humans, like those for the animals, are on either side of the room.

A skinny calf jogged awkwardly back and forth across the sale pen as the auctioneer began the thorough examination of its perceived worth.

"HerewegottanicelookingcalfstartingatthirtyfivethirtyfivethirtyfivefortythemansaysfortydoIhearforty fivefortyfivefiftycanIgetfiftyonthisnicecalfhere?fortyfivefortyfivefortySOLDtothebigbluehat."

The exchanged property trotted gratefully through the revealed exit as another, fatter calf entered the other side. Scott had positioned himself halfway up the center set of steps.

"WelllookatthisfatcalfIgotitweighinginateightypoundsgonnastartthebiddingatseventyseventyseventy eightyyeightyfiveninetyninetyyouheretotakemypicture?I'llprobablybreakthecameraninetyfiveninetyfive onceninetyfivetwiceSOLDforninetyfivedollar," concluded the auctioneer, waiting for an answer.

PLATE 2.11

Scott looked up.

"We're doing a book on small-scale food in the Valley, and everyone said we should check out Escalon," Scott said.

"You'regonnawanttotalktomybrotheroutside juststraighttothelefttherelet'sstartatfiftyforthis Holsteinherefiftydolhearfiftyfive?fiftyfive? fortyfivefortyfivefortysixfortysixseven..."

ON FRIDAYS THERE ARE TWO AUCTIONS at the Escalon market. The main building exchanges a standard selection of livestock—goats, pigs, calves, and sheep—mostly to men of various colors in cowboy hats or baseball caps, while an open-air market deals in smaller or more unusual animals, such as chickens, ducks, doves, pheasants, button quail, rabbits, canaries, guineas, peacocks, and, when we were there, one piteous tortoiseshell cat who had obviously done something very, very bad one too many times. The crowd outside is also different, composed of many more women and a larger range of ethnicities. Most, but certainly not all, of the animals sold here are eaten.

The auctioneer outside spoke a great deal slower than his more professional brother, included jokes, and often cajoled regular bidders into buying or paying more for an animal, noting, for instance, that these ducks came with the cage and that twenty-five dollars is really a very good price.

At the conclusion of the bidding, the crowd dispersed, some going inside to watch the larger animals, and others to pull their trucks around so that they could load up their purchases and go home. The Southeast Asians often preferred bags to cages, and I saw one woman stuff five ducks, stretched out into feathered torpedoes, into a mesh bag that had once held oranges.

We were looking at some rabbits when the twins, Brian and Brandon, came up and asked us what the camera was for. We explained and asked if they had bought or sold anything at auction. Little did we know that we were talking to the redneck princes of California.

"We raise meat rabbits. Those are ours over there. We sold them all today," said one of the twins.

"Do you skin and clean the rabbits too?" asked Scott.

"Oh hell yes," said the other twin. "I can kill, skin, and gut a rabbit in under two minutes. You just take a blunt object and hit the rabbit in the back of the head..."

"Make sure you hold the rabbit away from you, though, because when you hit one it will squirt a lot of blood right out its nose," interjected the first twin.

"Yeah. And then take your knife and cut a slit down its chest."

Little did we know that we were talking to the redneck princes of California.

"Make sure you don't cut too deep, or you'll puncture the guts and ruin the meat."

"Then you just rip the skin off. It comes off just like a jacket. After that, you pull out the guts, and that's it. If you do it right, the whole process should happen fast enough that you can feel the heart beat twice in your hand."

The twins showed us around the market and told us some stories, some of which may have even been true, of how fast their trucks go, the time they attached a Jeep with a tow rope to a truck and went mudding, and, most incredibly, the time Brian killed a bear with a seven-inch blade he always keeps upon his person.

"We were up on some ranch land in the foothills. A friend of ours owns it and lets us bow hunt for deer there. Well, I was up on a tree branch waiting for this buck I'd seen earlier, when I see Brandon running as fast as he can downhill. I ask him what he's running for, but before he answers I see this bear hauling ass after him."

PLATE 2.12

PLATE 2.13

"I was running downhill because four-legged animals can't run very fast that way. They trip up and start tumbling. That's the only way you can outrun a bear."

"Well, I see this bear chasing my brother, but before I can do anything, the branch breaks! I drop my bow and fall directly on top of the bear. The bear just keeps running, and what can I do? So I pull out my knife and start cutting into the bear's throat. It felt like I was cutting forever. But it worked. The bear fell over and there I was, sitting on top of a dead bear, covered in blood. So I pulled out my cell phone and called my dad. 'Dad,' I said, 'bring the quad, I'm sitting on top of a bear and we need to bring it down.'"

"The bear must have weighed six-hundred..."

"Four-hundred..."

"Since we were bor..."

"Since before we were born. We've always been Bubbas," finished Brian.

"These boys," I said to Scott after they had left, "have it good."

"Oh, and don't they know it," he answered.

We did not linger around the auction long after the twins went hunting. The bidding had begun on the pigs, and the air was rent by their angry, obstinate, and disbelieving screams. Calves are pliant, limp creatures, desperate to do what pleases. Goats and sheep are skittish, but predictable in their movements and easily controlled with a well-trained Australian shepherd, but pigs are individuals. A pig understands what you want from it and seeks the opposite.

The twins showed us around the market and told us some stories, some of which may have even been true.

"Well, probably two hundred pounds. My dad had to sit on the front of the quad to balance out the weight."

"We didn't have a license for bear, so we couldn't take it anywhere. We just dug a pit in our friend's yard, got some coals going, called a few people, and had ourselves a feast."

The twins spotted two bosomy blond girls down in the parking lot and suddenly became less loquacious, subsumed instead in the intense and wordless lust of all adolescent boys. These girls were not twins, explained the twins, because they had been born on different days. Before they left us to pursue their twin-on-not-quite-twin fantasies, Scott asked them what they were ethnically.

"Oh, we've got Irish, German, Portuguese, some Italian...," said Brian.

"But most people just call us 'Bubbas.' We've been Bubbas since we were six..."

"Four..."

Pigs will often charge their handlers, or, borrowing a trick of civil disobedience, just go limp and lie down. In those loathsome industrial pig farms, a fate all of these pigs had escaped, the pigs will riot, sometimes attempting successful jailbreaks. The pigs here were not being abused. They would all lead comparatively happy lives. But these pigs knew that obedience is the death of the soul. So they screamed. My head was pounding and there was no escaping the relentless Valley sun. It was time to go.

The fate of no animal was altered by my presence at the Escalon market, and I was discontent. I do not disapprove of the existence of livestock, far from it, but I do want to be able to, as my mother still does, change the fate of a single being from time to time. It was not the market or the protests of pigs too smart for their own good that dulled my mind with pain. I was angry, instead, at my apartment. What was I doing living in a place that would not tolerate and could not support the life of something so simple and basic as a single goat?

PLATE 2.14

Plate 2.15

Red, White, and Roan

During the tenderest of my childhood years, my mom milked cows at a local dairy. The small herd of Holstein heifers were milked twice daily, once at four in the morning, and again at four in the afternoon. This was only one of her three jobs, in addition to a paper route, and so very often when her alarm screeched at three in the morning, though it shook the house, and though it rested on the coffee table next to the couch where she slept, she did not hear it. I, an ethereal sleeper until I discovered alcohol much later in life, would ascend the concrete stairs from the basement where I slept, turn off the hateful alarm, and whisper *Mom*, whereupon she would wake up with a start and ask me what was wrong.

It pleased us both that she would occasionally take me with her to work, where my chief duty was to bring the milch cows in from the barn and fields where it seemed like they did nothing but eat and wait to be milked. The ladies knew the routine and enjoyed the relief that milking brings, so that usually just the appearance of a shouting five-year-old boy in the four o'clock dark was enough to set them into lines along well-worn paths to the milk barn, where my mom was waiting in a wet concrete pit surrounded by twenty-gallon glass milking jars. Sometimes one of the heifers needed to rebel, and I had to chase her around the barn, shouting, waving my stick, and trying to anticipate her next move until she gave up and trotted up late to the barn. I also bottle-fed the calves, but I found this to be difficult, sticky work. A free, unfrightened calf is a terribly irrational creature that refuses to hold still and will lick anything, including and perhaps especially five-year-old boys. My hair still hasn't settled down.

At the end of the shift, the dairyman would come down to the milking barn and ask me how my girlfriends were, a question that I found both condescending and uncouth. Condescending because I was one of those precocious children who found in nearly every question an exasperating denial of my true worth and abilities, and uncouth because it seemed like all you had to do was climb an apple tree with a neighbor girl and suddenly she thought she was engaged, and lost interest in the higher branches. Then the dairyman would fill up a five-gallon glass jar of fresh milk for us to take home, give us a wheel of cheddar cheese wrapped in red wax, and sometimes frozen paper packets of ground beef. I watched the cream rise into a soft crust at the top of the jar as it sat in my lap on the ride home. If the milking went quickly enough, there would be time to pour out a glass of milk, my arms shaking to control the jar, and spoon out thick dollops of cream before the bus picked me up for school, where nothing of any interest ever happened to me.

PLATE 2.16

And so, when Scott and I pulled into the driveway of Stuart and Emily Rowe, owners of a famous herd of Milking Shorthorns near Dixon, I was wary with memory. I had not been to a dairy since I was ten years old.

The Rowes are the proud keepers of the Innisfail Herd, the oldest continuous herd of cattle in California, which dates back to 1917. The herd is only just over two hundred head, which is minuscule by today's dairy standards, but is world-renowned for its quality and consistent performance in state and national competitions since the 1930s. In a couple of years, Stuart Rowe will sell the herd that has been in his family since 1924 to Rich Collins, owner of an endive empire, who is building a model working farm not far from the Rowe Dairy. Stuart will be retained as an advisor and he will still get to see the cows as often as he likes.

The Milking Shorthorn was once one of the most popular types of milking cattle, beloved for their sweet temperament, easy births, and handsome coloring, which ranges from red to white to roan. They were first developed in England and in 1783 made their way across the Atlantic to America, where they were prized for their milk, meat, and power.

Versatility fell out of favor in the twentieth century. Now almost all milk is produced by the Holstein, the famous black-and-white cow featured in commercials, which, although of unpredictable temperament, difficult birthing, and poor-quality meat, can produce nearly twice as much milk as most other breeds. Now there are probably fewer than three hundred small herds of Shorthorns left in the United States.

Stuart was having something of a midday nap when we arrived, and so Emily showed us around the farmhouse, a delightful dwelling with large, airy rooms and dark, heavy furniture.

"That's Stuart's first ribbon," Emily said, pointing to a wide banner hanging in one corner of a room covered in ribbons. "It was in 1939."

Stuart arrived and took us out to the dairy, roughly three miles from their home. He employs two Latino men who do most of the work on the dairy and who are provided with free on-site housing in addition to their wages.

"We'll give them calves sometimes, too. They raise them here, and then they like to slaughter them when they are about eight or ten months old. They invite all their family and friends out and have a huge barbecue. They'll have music, and dancing. They'll usually make sure we get

> The Rowes are the proud keepers of the Innisfail Herd, the oldest continuous herd of cattle in California.

a plate of food. It's a little, well, more flavorful than most of what we eat, but we've come to like it. We've had one guy out here for twenty years. He's bought an apartment in Sacramento to retire to. The other guy is completely different. He's bought a ranch in Mexico, but all his kids were raised here, and they don't want to go back. One of his daughters just had a little boy, though. He never had a son—something tells me he'll never wind up living on that ranch of his. The Mexicans," he went on, "are much more social than most Americans. They have these huge, tight families...sometimes I think they know how to get more out of life than the rest of us."

I asked him if he gave his workers any milk.

"The regulations and milk companies don't like that. We used to be able to give them milk, and now, well, we can't watch them all the time. Who knows what happens?" Stuart said slyly.

"If we were allowed to, we could sell almost all our milk to our neighbors. People still want fresh milk, but you can't get it anywhere. I don't know why fresh milk is treated like poison in this country," Emily, the more outspoken of the pair, interjected.

Stuart and Emily walked us to the milking shed.

"My mom milked cows when I was growing up," I told them.

"Did she?" asked Emily with a note of surprise. "I've always said that women make the best milkers."

The Rowes' milk barn is much like any milk barn, but theirs has smaller, plastic containers for the milk after it is drawn from the udders, instead of the beautiful glass jars I remembered.

Styrofoam cups and Emily, who carried a camera.

"You boys are going to try the milk? I've got to get a picture of that. Now, if anyone asks about the picture, you'll just say you were drinking coffee, right?"

"Oh, yes," we promised.

Stuart filled our cups from the pipe that pours the milk into the storage tank to prevent any possible contamination. I held my cup of milk tenderly, took a tiny sip, and was stripped of twenty years. The milk was complex, with layers of flavor and a crisp, sweet finish. I felt like a mammal again, bound, as we all are, to this sweet mystery of motherhood. There is some wilderness in fresh milk pulled still warm from living creatures, some ancient animal mystery that we, rather than understand, pasteurize and trap in plastic

---

"People still want fresh milk, but you can't get it anywhere. I don't know why fresh milk is treated like poison in this country."

The milking began and it was nearly time for us to head out to our next appointment. But first, I had something important to ask.

"Stuart," I asked, "is there any way we could get a glass of milk? I haven't had fresh milk since I was a kid..."

"Well, we're really not supposed to. Aren't you afraid you'll get sick?"

"No. Not at all."

"Well, let me just go get you both some cups," he said, and then went into the office. He returned with two

bottles. There are three hundred million people in the United States, and all of them are prevented by law from drinking fresh milk. But I pushed all this out of my mind in order to enjoy this, my last cup of fresh milk in perhaps another fifteen years. Neither whiskey nor wine was ever sipped so slowly, at least not by me.

Emily asked us to stand next to the sign for their dairy, and we smiled for her, our upper lips wet with the milk from some anonymous Milking Shorthorn, as the shutter snapped.

PLATE 2.17

Plate 3.1

THREE: Fish

PLATE 3.2

River Rats Want Sugar in Their Coffee

BOTH THE SACRAMENTO AND SAN JOAQUIN RIVERS empty into the San Francisco Bay. The arena of their confluence is an enormous man-made network of waterways, sloughs, marinas, and dikes referred to simply as "the Delta." Small communities of houseboats gather here and there and a few small towns quietly decline along the more commonly traveled waterways. Weekenders and day-trippers, fishermen or boaters on high-hulled luxury yachts or speedboats towing water-skiers, often descend on the eleven hundred miles of water in such numbers as to endanger one another. Keeping mostly to the more hidden sloughs, shanty-dwellers have, on small islands or afloat on the murky water itself, lived in seclusion and autonomy for over a century. The abundance of fish and the Delta's twisted enormity have made this life possible. We asked Steve Adams of S & C Guide Service, our fishing guide, to take us to one of these shanties. We had seen what it was to live off the land, and now we were curious to see what is was to live off the water, or, as our time was limited, we were curious to see a performance from such a life.

"I heard that the government is trying to root out the Delta shanty-dwellers," I said to Steve as Bill Conner's place edged into view.

"Yep. And one day they'll succeed."

The structure exuded a shabby yet perfect majesty. Thick, tall, dark posts rose calmly up from the water and through a deck, only to be topped by a thick roof of thatched palm fronds. Blue barrel halves lined the front of the dwelling, filled with various plants and small trees. Paper lanterns swayed from the rafters, a hammock bulged fatly between two posts, and an enormous banner advocated for Richard Pombo in the upcoming election. A variety of boats were moored along the side, many permanently. As we rounded the back of Bill Conner's home we saw the man himself talking with a bearded visitor in a small boat. As we approached, a feral five-year-old girl, barefoot, nimble, and chased by long ribbons of dirty blond hair, sprinted along a hodgepodge of planking and then leaped fearless as a kitten into her father's nearby boat, vanishing into its modest depths and emitting giggles. *My God*, I prayed helplessly, *never let this girl see the inside of a school building.*

"I picked up a whole slew of two-by-fours for you when you were gone. I saw them floating past in the morning. I'll bring them over," Bill's neighbor said as he started his engine and headed towards a palatial riverboat we'd passed only minutes before.

We hailed Bill and asked permission to come aboard.

> "I have a hatch next to my bed so I can fish in the winter without even getting up."

"Sure. I've got some great lies to tell," he shouted back at us. Steve had no interest in visiting with this legendary character, so he started casting around the edges of Bill's manse and told us to signal when we were done.

Bill enfolded my hand with his enormous mitt and shook it firmly while I tried to take the measure of the man. He stood well over six feet tall and was broad and barrel-chested, a hoary, bipedal ox. He wore only a torn pair of swimming trunks and the sun had reddened his skin and lined it with intersecting wrinkles that resembled enormous scales. Large, square glasses and a full head of thick white hair. His chest too was covered with curly, strong, pure white hair. His loose, old-man's skin rippled strangely when he moved, showing the formidable muscles underneath. He spoke in a thick, rich baritone hammy with mischief out of a jaunty lantern jaw.

"Let me show you around," Bill said. "But I always make my guests go first so I know where to step."

We passed two small houseboats, one filled with weights, exercise equipment, and punching bags, the other with a stove, a bed, and several musical instruments, and then climbed a ladder that led to an upper deck, where he kept a cistern for drinking water, a generator, a toolshed, and most of his plants.

Steve circled the shanty and fished impatiently. We had so little time. We would have to come back again later. I tried to get the basics out of the way.

"I used to own a party island," Bill said, "but I sold it a few years ago and now I just live here. I have a girlfriend in Fremont. I see her once a week and do naughty things to her. Then she makes me take a shower and sends me home. Over there is where I feed my raccoon," Bill said.

"Do you catch fish and eat it very often?" Scott asked.

"Yeah. Probably three or four times a week. I used to grow my own vegetables. Tomatoes, peas, cucumbers, but it's too much of a hassle these days and food is so cheap. I have a hatch next to my bed so I can fish in the winter without even getting up."

"Can I see it?" I asked.

"Sure," Bill said and then led us back towards the houseboat. "That hatch is good for other things too. The women like to wash up afterwards in it. Bigger women will sit on the edge, but little women I just hold onto under their arms and dunk them right in."

Bill lifted up the hatch and showed us the green water shimmering below. A pair of swallows had built a nest and filled it with eggs.

We asked Bill if we could come back another time and have him catch and prepare a catfish for us, and he gladly agreed.

"It's a shame old Hal Schell is sick. You'd really get a kick talking to him. He's written books about the Delta, too."

Before we could get out the door and signal Steve that we were ready to go, Bill stopped us with a miniature accordion in hand and began to sing and play a child's song with a child's enthusiasm.

"Twi...Twin-kle, twin-kle lit-tle star...how I...how I wonder..."

I looked to the floor. And I blushed.

WHEN NEXT WE SAW BILL he was somewhat subdued. As he pulled into the marina, we could see that he'd dressed up for us: he'd put on a shirt and was wearing an intact pair of shoes. I suspected that our formality had thrown him off a bit, and that if we'd shown up with a liter of whiskey and a

PLATE 3.3

PLATE 3.4

few chicks he'd have been more at ease—though for all his rhetorical filth I'd guess his hard living was behind him.

Bill was preoccupied with a pile driver he'd rescued at great effort and some expense and planned to use for some renovations.

"When I got it, it was covered in wild blueberries. You didn't ever used to see blueberries in the Delta. I picked off all the berries and tried to pot a few of them, but none of them took. I go around looking for nice plants a lot. When I find one, I'll dig it out and pot it. A man needs plants," Bill said, looking wistful and certain. "Without plants, you just don't belong."

When we arrived at his home he confessed to some domestic troubles. This weekend, owing to a doctor's

and lowered the hook, the line weighted with an old spark plug, into the Delta water. They stole his bait once and he had me retrieve another hot dog, but the second bite came almost instantly and it stuck. Bill reeled in his catch and frowned.

"Well, that's just a little one. I know there are bigger fish down there," he said, but Scott and I assured him that this would do. Certainly, this fish would do. A bigger fish, we convinced him, would have been too much.

Bill took the catfish downstairs, outside his houseboat, and began to clean it. Holding the head of this fish firmly in the vicelike grip of his first and middle finger, he made a quick circular cut and took off the head. Then, using a tool resembling an oversize pair of toenail clippers, he stripped

---

The flesh was moist and tasted faintly, and sweetly, of water itself.

---

appointment, he had stayed with his girlfriend for two days. Upon his return he found his shop torn apart and an indelicate message from his raccoon.

"She'll take me being gone for one day, but she punishes me if I'm gone for two," he said, shaking his head, it seemed, at women everywhere of every species.

We have strange occupations, Scott and I. There we were, visiting Bill Conner, a paragon of the Delta, asking impertinent and leading questions, sticking the nose of a camera into his intimate nooks and crannies, and then telling the man to fish for us. But he was more than game. We seemed to be useful to him in some way as well, but I hadn't yet figured out what that use was.

"I know there's catfish down there, big ones, because sometimes I feed them and they come up to the surface. Here we go," he continued setting a hunk of hot dog on an old hook. "I've had this fishing pole since I was a little boy."

Bill stood over a gap between his shop and his cistern

off the skin like a wet sock. Another incision took care of the guts, and we moved inside to prepare the fillets.

"These are miner's pans," Bill said, as he melted a tablespoon of margarine in a thin, much-used skillet. "They heat up really quickly, so I don't use so much gas. They don't make them here anymore. I had to get these in Mexico."

Bill cut four small fillets and threw them into the sizzling margarine, adding only salt and pepper.

"I don't keep much in the way of food or spices out here. I don't even use that sugar. I just keep it around because some river rats want sugar in their coffee."

One quick turn and three minutes later, catfish fillets were cooling on a plate. This dish is best eaten piping hot with your fingers. The flesh was moist and tasted faintly, and sweetly, of water itself. From catch to consumption had taken under eight minutes. Scott, who does not usually like catfish, made it through his fillet without a grimace and perhaps even some pleasure, and I went back for seconds.

After lunch our mood became more thoughtful. I asked Bill how his friend Hal Schell was doing.

"We just had Hal's funeral," Bill replied slowly. "It was quite a turnout. Hundreds of people were there. He used to wake up every morning at dawn, pull out his bugle, and play 'Taps,' so someone played a horn there. Afterwards, I went out with a few other friends and we scattered his

---

"I'm a symbol for something they've lost. It's in their eyes. But it makes them happy to know that someone is doing it."

---

ashes in the Delta. I hear they're planning a museum for him somewhere near the Antioch Bridge. No one knew the Delta like Schell. Do you two like poetry?"

We nodded or mumbled something, unsure of what was coming.

"As I walk the Delta sand,
I write my name...
and it is washed away.
Only, there is a record
with he who holds all
in his hands."

"I didn't write that. It's not even about the Delta. But I believe it to be true."

Bill then recited a poem from memory called "The Hyphen American," a rhythmic and sonorous plea for an American identity that needs no amending.

"Do you know who wrote that?" Bill asked when he was done.

We shook our heads.

"John Wayne."

Bill read to us a few more poems of a lighter nature out of a thick anthology called *Best Loved Poems of the American West*. In one, three cowboys go to town, get properly sauced, and then meet, and best, the devil on their way back to camp. He finished with an old San Francisco poem called "The Hyde Street Grip." His voice was sonorous, full-feeling and practiced. I felt that he had been reading and reciting these poems for longer than I had been alive.

"You've got it pretty good out here, Bill. This is the kind of life most guys only dream about," Scott said as we were preparing to leave.

"Oh, I see these guys that pass by on their boats, I see how they look at me. I'm a symbol for something they've lost. It's in their eyes. But it makes them happy to know that someone is doing it, that it is possible to live like this, on the water, with no women—except my girlfriend up in Fremont who I see once a week. That's why I'm out here. I have everything I need."

There was something enviable in Bill Conner's life. His groceries were heartily supplemented by the Delta beneath his home. His home itself had been built from the boats and scrap materials of the region, and in younger days he had even gardened. He made his home complete with potted Delta plants and the daily visits of his temperamental raccoon and had, in the decades he spent there, become an icon of the Delta, receiving visitors like the king of a miniature kingdom. It is not the old age I would wish for, but I am glad of those who do become permanently adrift.

PLATE 3.5

Plate 3.6

A Day on the Delta

In the beginning was Alpha and in the end is Omega, but somewhere in between occurred Delta—which was nothing less than the arrival of Man himself and his breakthrough into the daylight of language and consciousness and knowing, of happiness and sadness, of being with and being alone, of being right and of being wrong, of being himself and not himself, and of being at home and being a stranger. —*Walker Percy*

As we drifted from reed bank to reed bank in the ease and comfort of Steve Adams's bass boat, Scott and I became aware that we were seeing the Delta from a rarefied perspective. Boat owners make up a kind of gentry of the Delta waters, but along the banks of the sloughs another, more varied class of people make more democratic use of the murky shallows. Scott and I set aside one day to explore the maze of levees.

———————————

We saw him bent over a bank of blackberries along Highway 4 a few miles west of Stockton, picking berries with one hand and holding a jar and an old fishing rod. Big, kind, and simple, he was walking to his favorite fishing hole to catch supper for his mother. He picked ripe berries carefully and serenely past the thorns and spiders. We caught his picture and moved on.

This unobtrusive and modest rhythm marked most of our encounters for the day. The intricate webbing of the levees and our fundamental ignorance of their workings led to a day of pure photography. A father, son, and a dog fishing an irrigation ditch. Vans and trucks and cars parked along the side of pot-holed half-roads. Men fishing, children fishing, and women fishing. We saw barbecue pits artfully, and not so artfully, constructed from jagged pieces of levee concrete, black from recent fires. We saw a shack made from thatched reeds along the water's edge. Heaps of old furniture and appliances dumped in ugly piles. It was not a day of knowing, but of seeing. The only sure thing the writer can tell you about that day is this: when women go fishing on a Sunday afternoon in old, worn out, comfortable clothes, their hands red with fish guts, they don't want to be in any photographs. Don't even ask.

PLATE 3.7

PLATE 3.8

FISH • 109

Plate 3.9

That's Why They Call It Fishing

We met Steve Adams at Bethel Island around seven in the morning. We didn't realize it at the time, but this was a late start for our fishing guide, a bit of kindness on his part for two supposed city boys fishing the Sacramento Delta for the first time. We took a wrong turn or three before we finally found him, and then he didn't say much.

"Well. You two ready to catch some fish?" he asked, throwing down his cigarette and walking away from us towards the water.

We made early morning noises of affirmation and followed him towards his custom bass boat. In no time we were politely sliding by houseboats, moored fishing and pleasure boats, and a variety of signs in the genre of A DRINKING COMMUNITY WITH A FISHING PROBLEM, making towards a distant bank of tule reeds that marked the entrance to one of Steve's secret sloughs. A short biography of each of the three fishermen is in order at this point.

William caught his first fish at age five, a fair-sized bluegill in an overgrown cattle pond behind the Red Campbells' mobile home, with a stick, a length of string, and some government cheese. When informed of his options, to kill and eat the fish or turn it loose, young William demanded to keep the fish alive in a bucket as his pet. The bluegill died in that bucket, as the Campbells had said it would, a few short hours later. William never caught another fish, had never been in any boat smaller than a ferry, and remains haunted by the moral turpitude of his fishing past.

Scott is a once and future trout fisherman. Raised by a hunting and fishing father near the mountains of Washington State, Scott experienced moments of the wild American boyhood that has been praised in literature and song for three centuries. Those days are distant now and he tries to reclaim them in his own way, as his own man, through photography. He's ready to catch some fish.

Steve is the bass pro. After returning from Vietnam, Steve was taught to fish by a good friend and an expert fisherman. Once he got on the water he never wanted to leave. Now he owns two custom-built boats, runs one of the best guide services in California, owns a bass tournament, and has fished all over the world. He never says more than he has to, believes that if customers pay to catch fish, they should do so, and takes nothing but a thermos of black coffee and several packs of cigarettes from morning until dinner. He believes in fishing. He believes in the Delta, and in his own way, he's a fisher of men.

We may have looked like any three men in a boat out for a lazy day away from our wives, but we were each caught

PLATE 3.10

up in a personal and private drama. The water begins to speak to you. Every reality, sink or swim, is negotiable. The known world is mere surface, but underneath is another world filled with strange, living treasure obtained only through luck and trickery. Fishermen live in the world of folktale, of myth. Every line cast might hook a legend. Is it any wonder fishermen drink so much?

There are catfish in the Delta, sturgeon, shad, and crayfish, but we were here for one fish and one fish alone, the mighty black bass. Fishing for bass is a lot like three-card monte. The mark reads the reeds or rocks or trio of white irrigation pipes and says to himself, *there. There's the black lady.* He casts, gives the bait a jiggle on the way down, waits a moment, then reels in the lure. The boat has drifted further along by then, the cards have been reshuffled, and the mark carefully chooses once more.

We fished the thin, tule-lined strip of dark water for a couple hours with nary a nibble and I lost four or five lures on the reeds. However, I had put a hook through none of my fingers nor either of my own, nor anyone else's, ears. I had avoided throwing the entire rod into the water. I thought I was doing pretty well.

"Come on, fish," Steve said impatiently. "If they're not in the reeds then they're on the rocks. Let's go."

Steve restarted the main engine and before long we'd left the reeds behind and were on the edge of a bay, fishing along dikes built with rubbled concrete and covered with lizards. The sun was high above our heads in a cloudless sky and burned everything it touched. Near a strange and terrible feral fig tree, the only flora of any size on the concrete, I pulled my line in and contemplated the perfection of the Antioch Bridge. Nowhere is the clash of cultures in California as visible or as still. It is a cresting bridge, built like a drawn bow, and at its apex the unbroken stucco tan and concrete gray of Richmond, Pinole, Hercules, Martinez, Concord, Pittsburg, and Antioch suddenly vanish.

On the other side is a green and quiet landscape where small pastures nourish herds of sheep and cattle, and the ways of water rule the land. The fulcrum of the bridge, or perhaps the arrow in its bow, is a monument to the architecture of the apocalypse: a power plant. Closed scorpion legs of pipes and girding. A lone, tubular, tapering smokestack thrust like a stinger into the sky.

From below, from perhaps two feet above sea level, the tableau filled me with melancholy. A driver has little time to contemplate the scene. It moves towards him and away from him at the same speed. The illusion is that the earth moves beneath us yet we stay still. But on the water the bridge was still and I was still and the distant sheep were still and the cars moved but meaninglessly. The water told incessantly of impermanence, every wave against the boat there and gone, and suddenly the bridge, the refinery, the fields, and Antioch, though all stood bare and open before me, seemed like some distant memory or a madman's raving. How could I have ever thought it was real?

The fish weren't biting along the rocks either and after seven torpid hours, Steve called it a day. We'd try again the next morning on the Stockton Delta.

"Trout fishermen," Steve mocked. "You guys have no idea."

Scott essayed to defend himself and his kind, but I wasn't listening. Not only was it the fourth such exchange of the day, but I had interrogated my luck and knew it was changing. My cast had improved three or four hundred percent. The language on the water had changed as well. The question between boats and the shore, asked with two raised hands and a hunch of the shoulders, rather than being met with large, slow, sad swings of the head had recently been met with excited thumbs and even an occasional *this big* and a shit-eating, luminous grin. There were fish in these waters. Steve had even caught three or four, though only one was really worth keeping. The next one was mine. Scott didn't know it, Steve didn't know it, but I did. We were drifting along a series of tule islands when I saw the spot. *There she is*, I thought, *there's the black lady*. Neither Scott nor Steve had chosen this particular islet. My cast was right on target. Suddenly the line went tight and I gave it a quick jerk to set the hook, just as Steve had instructed. Steve and Scott put down their rods and came to my side.

"Pull 'er in! Keep your rod low to the water! That's it. No! Not like that!" Steve coached, his hands ready to grab my rod if my courage failed. Scott had the net ready. The fish leaped into the air, black and glistening and looking like the biggest thing I had ever seen.

sentiment. I had caught this fish with the express purpose of having someone cook it for me and then eating it. Steve had promised to take us to see a guy later in the day for just this reason. This was a fish of exceptional character, clearly, a black bass of distinction, but his luck had run out, mine had run in, and he was destined to be first lunch and then prose.

My luck, however, turned mercurial. Steve's main engine stopped. His other engine was a tiny, nearly silent propeller to be used while fishing, with a maximum speed of maybe two miles per hour.

"Keep fishing," Steve ordered while he messed with the engine and we puttered in the direction of the distant

---

The known world is mere surface, but underneath is another world filled with strange, living treasure obtained only through luck and trickery.

"Did you see that?" Steve asked. "That's a fish."

On the next jump we had it. It had taken twenty years, but I had caught my second fish. The bastard cut me as I removed the hook from his hard, thin mouth, but I didn't care. The fish flexed back and forth like one long muscle, its cavernous mouth opening and shutting angrily. This was no guppy, I realized, this was a killing machine, a lion, a king of his domain. He (she?) was uncomfortable, perhaps even in pain, but he was not frightened. This black bass was full of wrath.

"Well, don't just look at it, get it in the tank." Steve broke into my reverie. I put the fish into our storage tank with the others and then looked at Scott.

"Hey," I said, "weren't you supposed to be taking pictures?"

I had, I admit, fallen a little bit in love with my fish. *My* fish. Nobody else's. But I was a man of ideals now, not

marina. The hours passed. The sun dropped lower in the sky. We didn't catch any more fish. Finally we met up with the main slough that led back to the marina. Just as a passing boat offered us a tow, Steve tried the engine one more time and it started, albeit with a perpetual issuance of black smoke, and we limped, as they say, into port. Steve was wounded, proud, and worried. I asked if he could show us how to get to the place where we might get my fish cooked but he shook his head.

"There are eleven hundred miles of Delta. I only know how to get to the place on the water. I don't know how to get there on the road."

There was no point in keeping the fish anymore. He'd live to hunt another day, and I think all three of us were secretly relieved. I pulled him out, hooked his mouth with my thumb, and posed for my picture.

"Don't keep him out of the water too long," Steve said.

Plate 3.11

PLATE 3.12

Plish. There he goes…gone.

We offered our condolences about his engine and our best hopes that the problem would be easy and cheap to repair, but Steve was done talking for the day. He told us to call him when the salmon started running and then unloaded his boat in a dreadful, heavy silence. When we could watch no longer, we drove away.

THE SALMON WERE RUNNING, sort of. We had a five-thirty a.m. call at a gas station in Rio Vista, so we decided to spend the night in Isleton at the Rio Hotel. The hotel only serves its famous prime rib on Fridays, and Isleton Joe's was out of crayfish again, so we dined at Pineapple.

> You get the feeling that
> not much makes her happy.
> And not much makes her sad.
>
> But a lot of things
> make her tired.
> And a lot of things
> make her laugh.
>
> —Scott Squire, Isleton, California

He said this of our waitress at Pineapple, the old Chinese restaurant in old Chinatown in Isleton, the night before we met Steve and his boat, *The Other Woman*, for our salmon expedition. She told us a couple stories, one about a group of fishermen who caught their limit early and brought the catch to her, saying they didn't want to stop fishing yet. They never came back, so she fed the fish to her family. And her family, oy! She has a little brother who refuses to marry his girlfriend because he doesn't want to move out.

"Why would he move out?" our waitress asked rhetorically. "I've been taking care of him for forty years!" she said, letting out another bright laugh.

Perhaps, we thought, we would catch our limit and smuggle some fish back to her. The bass escaped, but the salmon would surely acquiesce.

FISHING FOR SALMON, AN ACTIVITY which is distinct from *catching* salmon, is not particularly exciting. In the slow waters of the lower Sacramento River, it is done by letting out roughly forty feet of line and slowly heading upstream. In faster, more northerly waters, the boat simply tries to hold its position on the water while the current does the rest.

Steve took us up the river, letting us off at the Ryde Hotel and later at Locke. On the grounds of the sumptuous and luxuriant hotel I gleaned three green pears and set them among the soft lead weights Steve keeps on a shelf under

---

It had taken twenty years, but I had caught my second fish.

---

the rim of the boat. We passed the dilapidated remains of an abandoned cannery and my eyes devoured greedily the network of posts suspending it over the river, the broken panes of the windows, and the seductive darkness within them.

"Oh," I said, "to be eight years old again and to live near that cannery."

Steve and Scott peered sideways at me like I'd said something crazy. I guess not everyone grew up exploring playgrounds of industrial decay, but to me, the broken windows, the network of undergirding, the secrets of the long low rooms within seemed primed for childhood adventure.

To take my mind off the notable absence of salmon, I cast around *The Other Woman* for objects of interest. Steve kept what looked like an old ax handle in his boat and I began, with an absentminded and unexamined bloodlust, to play with it, tossing the club from hand to hand and slapping it against my palm. *Rap. Rap.*

"What's this?" I finally asked.

"That's for clubbing salmon. When we catch one and

get it on deck, we use that to hit it until it stops moving," Steve said.

That club had taken many lives. I began to anticipate our catch with a new fervor. I could do this, I told myself, I could club a salmon until it stopped moving. They were the spawning dead anyway. No salmon undertaking this arduous trek would return to the ocean alive. Nature had already condemned them. My hand would be swift, and, in its way, merciful. *Rap. Rap. Rap.*

———

AND THEN... WHEN WE LEAST EXPECTED IT...*At the very last hour of the day...*

But no. Suspense, the spice of every victorious fishing story (the only fishing story worth telling), is a trope I

two before they pulled up a thirty-inch female, bursting with eggs. They fell on that fish like a pack of wolves. The scientists cut the fish open right there on his deck and began scooping up the roe with their hands and pouring it into their mouths.

"They didn't wash it or anything," Steve said, with true horror in his voice. "I've seen a lot of things in my time, and I've got a pretty strong stomach, but watching that, I thought I was going to throw up. I really did."

In Red Bluff, where the Valley narrows to a thin point between two mountain ranges and is shaded on either side by deep and ancient forests, Steve summed up our adventures.

"Well," he said, "that's why they call it fishing."

———

The fisherman, and the hunter as well, require access to a commons,
a free realm of nature that is open, theoretically if not actually,
to the multitude and owned by none.

can no longer permit myself. We caught no salmon that day, nor on the next—our last—trip. Late and heavy spring rains had conspired with a series of nasty heat waves to delay the salmon's run. Salmon populations have been declining steadily since the precontact days of living myth, when, it is said, a man could cross a river on the backs of the fish. Dams, pollution, and river re-engineering—all had played their part. We sought salmon, as we later discovered, during the worst year on record. So, um...*Let's tear down those dams!*

Steve told us a story of better times. He'd taken out a group of four or five foreign scientists up near Red Bluff. They hadn't had the lines in the water but a minute or

PART OF OUR BAD LUCK could be blamed on the exclusivity of our intentions. Had I been thoroughly committed to the edible when I caught the black bass, I could have purchased some ice and a Styrofoam cooler and prepared the fish at home, or even in one of the makeshift fire pits that scar the levees all across the Delta. Had we been fishing for food, we would have brought some lures for the ubiquitous catfish and used them on the Sacramento River, but we were greedy for salmon and consequently left empty-handed.

Fishing varies greatly from farming and ranching as a means to produce food. The role of the individual is greatly diminished in many ways, yet tremendously enhanced in others. The fisherman, and the hunter as well, require access to a commons, a free realm of nature that is open, theoretically if not actually, to the multitude and owned by none. Because the beneficiaries of the commons all possess the same rights, special significance is given to knowledge—genuine, occult, or, most commonly, a mixture of both—pertaining to certain spots, techniques, equipment, and times of day. In farming and ranching, the personal is converted into the public, the acquired knowledge and aggregate of individual labor produces from private property peppers and eggplants and corn and eggs and makes them available to a public that is increasingly global. The hobby fisherman transfers the public, which can be as wide and far as the ocean, and makes it personal.

The Japanese have a theory of flavors that assigns each category of taste a specific aesthetic and spiritual force. For a complete experience, all five flavors should be accounted for. Perhaps it would be appropriate to begin charting the kinds of intimacy drawn from the provenance of our food, and to draw, then, a model of wholeness. That which we gather, from the commons, ourselves. That which is gleaned for us by loved ones, carried in the improvised basket of an untucked shirt. That which we grow first for others and then ourselves in a converted chicken house on the side of the road. That which takes hours to prepare, and that which can be eaten off the briar or, like the carnelian roe on Steve's boat, freshly spilled from a thirty-inch salmon in the pool where she was born.

Plate 4.1

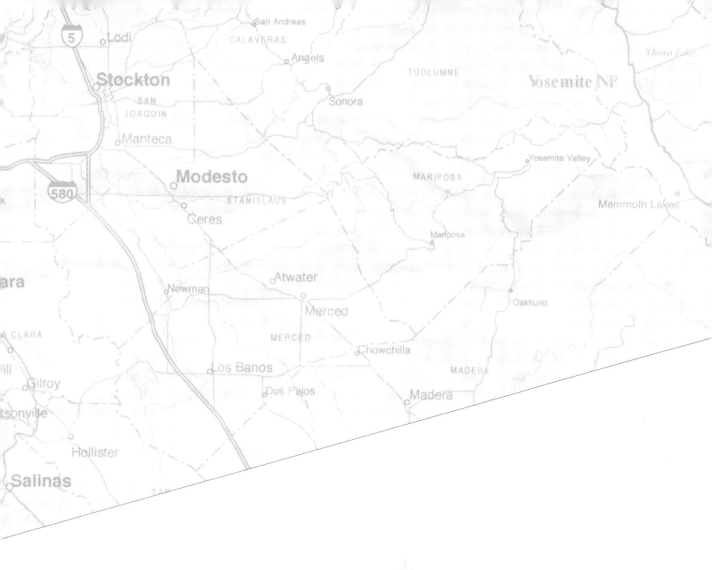

FOUR: The Kitchens

PLATE 4.2

Tucos Wine Bar and Café

IN THE MIDDLE OF OUR YEAR-LONG JOURNEY, Michael Pollan's book *The Omnivore's Dilemma* exploded with seemingly epochal force on the great waters of the Food Question. In the months that followed its publication, it seemed as though we had arrived in a post-Pollan universe. The book is mainly composed of adventures in other people's ideology, with Pollan attempting a temporarily objective entrance into various man-made food chains, from the fast-food machine to the glorious hunting of forest-fire mushrooms: a straightforward approach, but the timing and tone of the book coincided perfectly with the growing celebrity of its author to produce a seminal text. One of the most compelling conceits of the book is to follow an "ingredient" (in the terms of scientific reductionism) from its origins to a consumable setting, and we decided, as conversation and homage, to follow suit.

KEN LINDLEY'S LONG, LANKY ARMS were thrust out from his body in the posture of one crucified.

"Turkeys are actually very stupid," he said, and then broke the cross with a loud clap of his hands.

"Gobblegabblegooblegob," said the toms and hens gathered in an accidental huddle.

"It's really an involuntary response. They do that to any loud noise. They can't help it."

Ken and his wife, Deborah Raven-Lindley, own Nevermore Farm, outside of Arbuckle. Their eggs, tomatoes, peppers, and eggplants are routinely purchased by Prudencio, the owner of Tucos Wine Bar and Café, in Davis. Pru, with his wife, was on his first visit to Nevermore Farm and was there mainly to make the introductions. We would document the whole process, from egg, tomato, pepper, and eggplant gathering to egg, tomato, pepper, and eggplant preparation and consumption at Tucos. After that, the food was on its own.

Pru's daughter succeeded in laying her tiny hands on one of the Silkie chickens, and then Pru had to get back to his restaurant, leaving us to fetch the eggs and produce. They are busy people, these restaurateurs.

Deborah and Ken are still novices in the farm world, having only purchased this land and begun raising turkeys, chickens, and some vegetables on it a few years ago. They are also artists. Small, colored bottles hang from the branches of many trees, sculpture adorns their lawn, and their chicken coop is built to resemble a lighthouse. They are collectors as well, but not of conventional paintings or pottery. They collect living art. Almost every bird they own is an heirloom, or "heritage," variety, and they view their nine acres as a conservation farm. Industrial farming, it seems, has decreased our options not only of tomatoes, apples, and squash, but of the animals that feed us as well. Does a heritage turkey, say, a Narragansett, taste better than a cage-crippled and doped-up butterball? You can find out for yourself. Deborah and Ken began selling their turkeys for Thanksgiving dinner the year before our visit. They can be ordered a month in advance and will be freshly killed, cleaned, and shipped to you by the owners themselves.

Scott spent some time chasing the poultry while I investigated the lighthouse coop. A plump red hen sat with the utmost dignity in a box certainly containing a few eggs. Ken unceremoniously ruined her repose and

with our catch, and we started the leisurely walk back.

Suddenly a man on a backhoe appeared and dismounted.

"The jerky isn't ready yet!" he shouted.

"Dammit," I replied, sensing a routine of instant familiarity common to rural characters. "That's why we came. What do you mean the jerky's not ready?"

"I said," he repeated as he walked closer, "the jerky isn't ready. It needs a couple more days," and then he was before us, quick-moving, hale, and hearty-voiced Harry Sumsutch, Deborah and Ken's neighbor, friend, and mentor in the farming world. He walked with us as far as his house. On

---

"I find the produce first, then I make the menu and begin playing with the flavors. We're living plate to plate.

---

she left the coop indignant. The happy countenance of a full nest beamed up at us. Ken gathered the eggs in his hands one by one.

In order to retrieve the produce, we took a walk through the surrounding almond orchards. This season, Deborah and Ken were using a small piece of land belonging to a neighbor and friend, Harry Sumsutch. The land in which the garden was nestled had been stripped bare. A deep chasm ran through it. Harry, Deborah told us, was a conventional almond grower, but then again he wasn't all that conventional. The barren land we walked through was being transformed into a bee habitat. Irrigation tubes crisscrossed the dirt, and already isolated clumps of lavender had taken root. In two years, this land would be covered in flowers and the chasm would be a creek.

We waded through the knee-deep garden and fished for peppers, eggplants, and tomatoes, all heirloom varieties, of course. Two paper bags were soon bulging

the top of his hot tub, four screened boxes were propped up to the sun. Sure enough, the jerky wasn't ready.

WE TOOK OUR TEMPORARY JOBS as unpaid egg and produce deliverymen very seriously as we drove down I-5, marveling at another red Valley sunset. The earth exhaled so piteously during the almond harvest that the sky bled for it at dawn and dusk. The world was dark as we parked in downtown Davis, unloaded our precious cargo, and strutted, joyfully burdened, towards the convivial sounds and sights of Tucos Wine Bar and Café. ("Tucos" is how Pru's daughter pronounces the Portuguese word for juice.)

Megan, the waitress, greeted us cheerfully and bade us sit and be welcome, and we watched the syncopated music dance through the air. Tucos is a small place. The tables are small, the plates are small, the staff is small, but the wine selection, the flavors, and the hospitality are large. The prices are what the prices are, but who needs money,

PLATE 4.3

Plate 4.4

anyway? Our bags were inspected, appreciated, and taken in back. Two carafes of wine appeared, red for me, white for Scott. Pru walked past, talking with Megan about the baked cranberry beans, a new addition for the fall menu.

"These flavors are like a fireplace. They make you feel good."

Scott began negotiating with Bobby in the kitchen, saying things like "Is there a light in your oven?" and "Am I in your way?" as I sat down and talked food with Pru.

"I live in the Bay Area now," I began, "and we're all very proud of our markets and our restaurants, but if I try to talk about food in the Valley, people just stare at me. But I say that some of the best food, not just farmers' market produce, but restaurants, butchers, everything, is found in the Central Valley. That's the story we're telling."

"Exactly," Pru replied. "I keep telling everyone I see about the amazing food here. I just spent two hundred dollars on farmers' market greens yesterday, and they're the best I've ever had. When I opened this place, the food was mainly just to complement the wine, but as I started getting to know the farmers at the market, I realized what a charmed place this was. The flavors are just amazing. I barely even have to do anything to make my food delicious. It's all in the ingredients."

To illustrate the point, Scott appeared with a scallop ceviche with red cabbage, carrot, cilantro, red pepper, and onion. The crispness of the vegetables complemented the sweet, citrus-soft flesh of the scallops. The cabbage especially startled and pleased. Pru was eating a cachapa, which is a type of cornmeal pancake, and pouring sparkling water into a glass of strong red wine.

"I practically live off these when I'm in Venezuela."

Pru began his adult life as a computer programmer, but food and wine became a stronger calling as he traveled.

"I didn't use to even like wine. I would buy bottle after bottle, try a glass, and then pour it down the sink. Then I had a bottle of wine that really moved me. This wine was different. It was a Grgich Hills Cabernet, 1987. I understood, then, that wine could be, should be, a spiritual encounter. I began talking to wine merchants more, having them guide me, and I realized how important a wine education was. That's the main reason I opened this place."

Pru's commitment to wine education is evidenced by his wide variety of flights, each designed to show the range and character of wine from regions around the world. The wine selection is inventive and instructional, but the food makes a statement. Pru's commitment to fresh, local food and the abundance from which he is able to choose have produced a cuisine as dynamic as the Valley itself.

"This is a lot how I would cook at home. The menu is based on whatever I find at the market. I find the produce first, then I make the menu and begin playing with the flavors. We're living plate to plate."

Even though we had visited Nevermore Farms but an hour before eating an expertly prepared meal from the ingredients, when the food appeared before us its origins were just abstract bits of memory. Pru's personal vision of cuisine was exciting, and it was animated by the language of the unrepeatable and the new, but it was not just the flavor and freshness of the preparation that obscured the food's origins: there was also the luxurious candlelit wine somnolence of Tucos itself. Already, one level of intimacy had been removed. We were in a lower sphere of edibilism than we had previously haunted.

There are many wonderful restaurants in the Valley, but in them, unless you have a personal relationship with the place, the owners, or the employees, the set of virtues we've described as edibilism is noticeably absent or mere rhetoric. Sometimes, however, food is more than food. Food can be an arcade into a genuine tradition. Tradition, like food, can be a city of intimacies and, like cities, can be visited, can be explored in idleness or earnest, can be known.

PLATE 4.5

Sopash

JOHN PEREIRA HAS THE VISAGE AND BUILD of a middleweight boxer who won more than he lost. He is a hired gun on the California festa circuit, and in Buhach he reigned supreme over a low wall of simmering vats and a team of young men being initiated into the tradition. Raymond, Matt, Joey, and John T. had helped prepare the sopas before, and they pretended to buck and chafe under the unwavering and critical attention of the older men, who pretended themselves never to have seen a more troublesome and irresponsible group of boys in their long and venerable lives. The truth was that both generations were having a great time.

Owing to the large numbers of Portuguese immigrants all over California, there are over two hundred festas celebrated every year, the largest celebration occurring in Gustine. The festa, an explosion of food, community, and religious pageantry, is a cornerstone of Portuguese Catholic tradition, and sopas is its cornerstone. It is a traditional dish invented, it is said, by Saint Isabella herself, and used then as now to feed the hungry masses. A sopas dinner is open to all and is always completely—for the noblest of religious reasons—free. On this night, John Pereira and his crew would feed over one thousand people in Buhach.

A Portuguese church usually has a special kitchen installed to prepare this unique dish. The Buhach church devoted an entire wall to over a dozen enormous vats, each heated from below with a slow, even heat, and each containing a whole side of beef donated by local cattlemen and dairymen. The meat simmers for ten to twelve hours (some prefer to cook it even longer) in a rich broth made of sauterne, garlic, onion, tomato sauce, ketchup, bay leaves, allspice, black pepper, cumin, salt, and cinnamon. An hour before serving, halved heads of cabbage are dumped into the simmering broth. Meanwhile, thick slices of traditional bread are put into large stainless steel bowls with a generous spray of fresh mint. Once the cabbage is pliant, it is removed from the broth and set in a separate vat. Then the meat is removed from the bones, the bones being discarded and the meat reserved in yet another vat, leaving only the broth in the original. The broth is poured over the bread and mint, then some cabbage and beef are reintroduced. One bowl will easily feed six to eight large, hungry people.

The labors in Buhach were entirely segregated. The men prepared the sopas while the women attended to the beauty of the event, including St. Isabella's crown and the carriage and deportment of the queens. Each church was responsible for two queens, a child and an adolescent. A queen is chosen not because of her beauty or, as many other queens are, because of the number of tickets that her supporters have sold, but merely by being related to that year's presiding officers in the church community. The duties of a queen are simple. She dresses in a beautiful and ritual gown, piles up her hair, wears a crown, and attends every festa she can. A queen walks about, poses for pictures alone and with other queens, and generally looks pleasing and regal until her shoes start to pinch.

When we arrived, a group of older men were munching on tramoço beans. One very old fellow was slicing long loaf after long loaf of bread with a band saw. Another man heaped the bread into the stainless steel bowls, added the mint, and stacked them three high on a long steel table. John Pereira was supervising the boys as they prepared to

this ledge each member of the line sat in grand comfort, shaded, even, from the light of the sun by the building itself. I resolved then and there to include such seating along any buildings, save a barn, that I might design or construct. How gracious, how humane!

As the boys, wearing thick, heat-resistant, elbow-length gloves, were taking the bones and meat from the stock, a little blond girl who had taken a fancy to Scott, or at least to his Rolleiflex, screwed up her face in disgust.

"Why did they use all the bones?"

"Because," Scott answered, "that's where all the flavor hides."

The Buhach dining hall was separated from the kitchen by a counter and a series of removable partitions. The

---

Sopas...is a traditional dish invented, it is said, by Saint Isabella herself, and used then as now to feed the hungry masses.

---

dump the halved heads of cabbage. One man lifted the heavy lid of a vat while two others dumped the cabbage splashingly into the broth.

"Ooooh!" exclaimed the onlookers. The kitchen air became thick with the smell of cabbage, wine, and cinnamon.

While the cabbage cooked, Scott and I ducked outside. The appointment of new officers was taking place in the meeting room. We crept inside and stared admiringly at the walls adorned with photographs of past queens reaching back to the thirties. It was rumored that the oldest living former queen would be making an appearance. Once the official business started we fled.

Outside, people were already lined up. Scott did his reporter thing and made small talk with the more visually interesting members of the queue while I admired a first-rate example of civilized architecture. Built into the wall of the church meeting room ran a low ledge, and on

dining hall was long, narrow, and shallow. The walls were painted bone white and the ceilings were low and ridged. It looked like the belly of a whale set for fine dining. At the far end of the hall was a bar.

The tables were filled by an orderly and happy horde. The partitions had come down. The men had been at work since six in the morning. It was seven in the evening. Time to serve the sopas. Two men stood over the vat of cabbage, another stood over the beef. A bench had been placed near the vats of broth to function as an assembly line for moving the stainless steel bowls of bread and mint along. John Pereira wielded his ladle, a cast-iron saucepan affixed to a broom handle, like a quarterstaff. A thick layer of salt was poured over the floor to improve footing. Runners stood on the other side of the counter, ready to carry the sopas to the tables. Suddenly John ladled the first cataract of broth onto the bread, releasing a quick whiff of scalded mint, pressed the bread into the broth with the

PLATE 4.6

PLATE 4.7

Plate 4.8

PLATE 4.9

bottom of the second bowl, and passed it to the meat and cabbage men, and no one stopped moving for almost an hour as bowl after bowl was filled and passed along to the not quite ravening crowd beyond.

At this point, there was nothing for me to do but shuttle beer to Scott and try to catch some of the women whose business was beauty—not the queens, mere figureheads, but those behind the queens, quick-moving women with flashing eyes under dark eyebrows, brash and imperious in two languages. Apparently, and this is perhaps the most adorable and distracting thing ever, the final *s* at the end of Portuguese words is pronounced "sh." Thus: *sopash*.

"Where," I asked Janine, former queen and presiding secretary, "do all the queens get their dresses and why do they match all the decorations?"

"There's this old Portuguese woman in Los Angeles who does it all. The whole second floor of her house is filled with this stuff. When a queen is chosen, she goes down to L.A. and gets her dress made and buys her crown. She makes everything herself."

I talked to the former president, who told me more about their patron Saint Isabella.

"She was a queen in the 1200s. Her husband and her son had a falling-out, and the son left the castle. He returned years later with an army, determined to make war. The two armies came together, one led by the father and one led by the son. Just before the attack, Queen Isabella rode out in between the armies. She stopped halfway between father and son and a dove landed on her hand. Father and son both dismounted, took off their helmets, and embraced. She is a peacemaker, a healer, and a friend to the poor. She's the reason we do all of this," he said, gesturing to the furious activity in the kitchen, to the jovial noshing along the long tables, to the very existence and design of the kitchen and dining hall itself. "We keep her crown over there. We allow members of the church who donate money to the church or demonstrate great need to keep the relic at their homes

No one stopped moving for almost an hour as bowl after bowl was filled and passed along to the not quite ravening crowd beyond.

for a week. People will gather at the home and pray with the owner, and then the crown moves to another home. None of the food goes to waste. If we have too much, we donate what is left to the food bank."

The kitchen began to slow down and the tables to empty. When everyone had been fed and fed well, it was time for the men in the kitchen to have their share. John Pereira set up three more bowls and grasped the handle of his ladle one more time for his portrait.

"Look at that technique!" the boys taunted as John poured broth over the last three bowls of bread and mint. "Beautiful form! I've never seen anyone pour like that! Bravo!"

John invited us to sit and eat with the staff as they took their meal at the nearest table. We piled beef, cabbage, and bread high on our plates and tucked in. The cabbage nourished and comforted, the meat was soft and luxuriant from hours of slow cooking, but the bread inhered divinity. As the broth passed through each pore in the bread, the myriad flavors were caught there and fixed. Cinnamon, bay, garlic, onion, and cumin, suggested in the meat and cabbage, blended seamlessly in the broth; in the bread each stood apart and sported distinctly with our taste buds. Such prodigious absorption of flavor came at a price. The bread was martyred in the broth, and slowly lost its solid state in the bowl, on our plates, in our spoons, and finally became a kind of porridge on our tongues.

"For not being Portuguese, you two really put away the sopas," John Pereira said, after we had picked through the communal bowl for our third personal helping.

"Uhnfrumnahumm," we replied in unison.

Plate 4.10

The Wool Growers Rest

THE DUCK HUNTERS CROWDING THE BAR of the Wool Growers Rest in Los Banos all wanted whiskey and water but we were here for some Old World sorrow. Michele, the bartender, was the keeper of an ancestral reticence and we were four beers and thirteen questions into a laborious but fascinating extraction of information. A Basque, and no young man, Michele runs the bar of this historic restaurant and onetime boardinghouse with his wife. His brother, Gabriel, is in the kitchen.

Michele and Gabriel bought the Wool Growers Rest in the seventies from the original owners. It is one of many like places scattered throughout the western states as a natural outgrowth from the large Basque migration. There is no menu and the tables are long. Once a stopgap cure for starvation and despair, the Basque restaurant/hotel is now a treasure of the West and an important glimpse into another world.

I was concerned, having just found this place and registered the venerable age of its owner, that, like so many good things, it would vanish. I asked Michele, who had almost wordlessly dispatched the thirst of the duck hunters, if the new generation was interested in carrying on the restaurant. Michele seemed not to hear the question and walked to the other end of the bar. He stood there for some time, staring, polishing glasses. I finished my beer. Michele returned, I ordered another beer, and he answered the question. We had grown accustomed to this ritual.

"No. These kids have all been to school and don't want to work. This is hard work. My brother, Gabriel, is in the kitchen at four in the morning, butchering the lamb and getting these huge pots going. We don't leave until after ten o'clock. This is hard work. But once a kid goes to school, he's ruined. They teach them not to work there. This new generation is not Basque. They are American. They don't understand work."

I understood that the word "work" meant something different to him than it did me. It was not mere labor. Work was pride, tradition, devotion, history, and family—all the big words—and this is what the schools had taken from his people's children.

"You butcher your own meat every morning?"

"Yes, the truck brings the animal whole and Gabriel prepares it in the traditional Basque way."

"Do you think we could go into the kitchen and have a look at these pots?"

Michele shocked us both by laughing, though even then he was careful not to smile.

"Sure. Just tell Gabriel I said you could. But watch out. He's not as friendly as me."

Michele returned to the far end of the bar for good. A bowl of soup, some bread, and a large plate of lamb stew appeared in the window that connected the bar to the dining hall. Michele retrieved it and began to eat. His every mouthful seemed to evoke a fierce nostalgia. This was not a meal. This was obeisance.

We left the bar and strode boldly into the hall. We found the entrance to the kitchen, and with wide grins, weak knees, and fast talk made our way past the bristling wait staff and through the heavy swinging door. It seemed much like any professional kitchen, complete with Latino help, but we didn't have time for a lengthy inspection. Gabriel, the younger and larger brother, was on us instantly. He did not address us. His glower spoke for him. He did

Less than two minutes later she reappeared with a large stainless steel bowl filled with a hearty vegetable soup, a ladle, a long plate of navy beans, two salads, a basket of bread with pats of butter, and a decanter of light red wine.

"The beans are for the soup," she said.

I tasted the soup first sans beans. Carrots, cabbage, leeks, thyme, parsley, turnip, onion, potato, and garlic merged seamlessly in a rich stock of chicken broth. My flesh tingled. Who was it, I thought, who loved me enough to send me such soup? Hesitantly, I added a spoonful of the navy beans. The natural gravy of the beans thickened the soup and lightened the color of the broth. I brought the spoon to my lips.

"Fuck."

Michele, the bartender, was the keeper of an ancestral reticence and we were four beers and thirteen questions into a laborious but fascinating extraction of information.

---

not want to resort to violence, it said, but we were forcing his hand. We began to stammer.

"Michele...pots...writing a book...said we could..."

"This is a small kitchen," he broke in with exasperated fury.

Chastened, we left the kitchen and sat down at the end of one of the long tables covered in red and white tablecloths and waited to see what would happen to us.

A vivacious woman in a traditional red outfit came up and smiled at us. Despite the costume, the smile gave her away. She was no Basque. Gabriel had found himself an American wife, a crocus in the snow.

"Pork, steak, lamb, or chicken?" was all she asked us.

"Steak," said Scott.

"Lamb," said I.

"Okay," she said, and walked away.

Scott looked at me. I looked at Scott. And we knew, from that moment, that this was to be one of the best meals of our lives. We looked again into the steel bowl. There was enough soup for four. I buttered some bread and got to work.

"What's this wine?" Scott asked when Gabriel's crocus-in-the-snow returned with a deep dish filled with lamb stew. The wine was light but had a mouth-cleansing sourness. A loyal servant to the food, it kept no secrets and plotted no coup. Gorgeous, for those who find beauty in modesty.

"It's made special for us. It's a lot like a Spanish table wine."

"And this dressing..." I asked with a mouthful of salad.

"Yes, that's house-made."

We regarded the stew. We hadn't finished the soup

PLATE 4.11

Plate 4.12

yet and I was already beginning to feel full. I hefted a chunk of lamb onto my fork and skewered a piece of potato and a bit of carrot. I ate the morsel in silence, pulled a small, clean bone out of my mouth, and looked away. I felt my face harden and grow distant. I had never before experienced terror in mercy. Was there no true thing that did not seek to immolate us? Strange, the cuisine that creates asceticism out of abundance. Some nascent understanding of these lonely, ancient shepherds was born within me. This food, I thought, in the very intensity of its comfort, presumes such a loneliness that I almost could not bear it. Instinctively, I reached for my wine. It cleared my head and returned me to myself.

Our waitress paused at our table. We stared at her with no little fear.

"Did you want any pigs' feet?" she asked, casually taking our measure, noticing the food left on our table and the glaze on our eyes.

I recalled my Olga, a beautiful, slight, Russian-Jewish girl who is a great gnawer of bones, an eater of chicken feet, of cold meat gelatin garnished with horseradish, of apple cores, and I said, as though it was a matter of principle, "Yes."

The pigs' feet are boiled until tender and served cold with a bright, homemade tomato sauce, fresh in the summer, canned for the rest of the year. I ate one with interest and a mixture of pleasure and disgust. Scott's response was mostly disgust. I marveled at how tangy and sharp the cold red sauce was, and how, despite the pleasing flavor of the feet, the chewy texture still took some self-control to endure. It was a perfect interlude, however, to try something that didn't frighten us with its perfection.

Which was good, because our main course had just arrived. Scott's steak luxuriated amidst thick French fries and my lamb shank had been braised in a dark, tart wine

sauce. These, too, were delicious, and we manfully set to finishing at least this course. Aside from Scott's fries, we succeeded, then returned to the soup and sopped it up with the remaining bread. The dangerous, soul-shattering lamb stew remained on the table. Our wine had long since run dry.

"Ready for some ice cream?" our waitress asked. One of us must have nodded, because two cups of vanilla ice cream and a bottle of chocolate syrup appeared before us. We ate that, too, and had our waitress put the stew in a bag for us to take home.

The stew eventually made it home with me and I warmed it for Olga. Our neighbor, Nesta, knocked on the door and we invited her in. Hospitality is our mutual rule

---

Who was it, I thought, who loved me enough to send me such soup?

---

and I began to tell her about the Wool Growers Rest and the stew when Olga said in a quick, strange voice: "I think it's gone bad. I don't think we should give Nesta any. It was in the car for a long time." Puzzled, I turned off the stove and offered some chocolate wafers and tea instead. After Nesta left, Olga disappeared into the kitchen. When I took the dishes in, she was halfway through the stew. Scandalized, I began to speak, but Olga cut me short with a strange look in her eyes. "What?" she growled. "It's mine."

Back in Los Banos, Scott and I ogled each other in gluttonous disbelief, unable to do else. Eventually, we realized, we would have to attempt to stand and walk out of this place, but, thank God, not at that moment. A man near us sat alone, his face dour, his eating patient and heavy. He had the same high forehead and hawk's nose as Michele and Gabriel, the same hard expression. Gabriel came out of the kitchen and they greeted each other very warmly in a language strange to us. Hands made timeless gestures and it looked as though jokes were being told. Then, abruptly, both men resumed silence, and their features again aspired to immobility. Gabriel returned to his kitchen, and the man to the toothy contemplation of his leg of lamb. How rare those moments must have been to the migrant Basque when these rests were established, when he could speak his own heart with his own tongue and know he'd be understood.

Finally, it was time to pay and to move on. We endured the teasing our waitress gave us about not finishing our meal, paid less than twenty dollars each, tipped well, and stumbled, groaning, out to the car.

Later in the trip, during a long, mistaken drive up Highway 245 through the Sequoia National Forest east of Fresno, I saw a number of thin, white waterfalls issuing from harsh cliffsides and thought again of Michele and Gabriel. The generous ebullience of the whitewater and the fierce crags from which it fell were not opposed to one another, but were necessary to each other. As this idea grew in my mind, I thought of the apparent disparity between the voluble excess and heartbreaking warmth of Basque food and the austerity of the people, and the disparity vanished. A hard, solitary people would create an elemental cuisine of undeniable comfort. They needed it more than the rest of us.

HuntFishCook

In the kitchen of Scott Leysath, better known as The Sporting Chef, the red thrill of the hunt and the burden of the kill are transformed into haute cuisine. Hunter, fisherman, restaurateur, chef, author, and television personality, Scott Leysath has made a life's work of this fleshy alchemy. Witness his Elk Tenderloin with Crispy Cornmeal Crepes and Fresh Papaya, or his Hearty Boar Soup, which combines a rich stock of smoked boar pieces with beans, leeks, carrots, zucchini, potatoes, garlic, and elbow macaroni. He's humble for a man who has made a comfortable living and even acquired a fair amount of fame participating in the greatest of sports in the world's most beautiful locales and then turning the results into kingly feasts for the entertainment and education of the masses.

"I just hope no one ever finds me out," he joked, when Scooter and I pointed out his good fortunes. (To avoid confusion, in this chapter, photographer Scott Squire will be called, against his will, "Scooter.")

Scott Leysath had just returned from the South Pacific, where he had been catching and cooking the enormous marlin. We oohed and ahhed in admiration but he waved the noises off. The joy of his homecoming overtook the joy of his adventures. He'd been traveling for some months now and his family, dogs, and home called out to him. We could see the relief in his face.

"Give me a nice Valley stream and some trout fishing any day. That's really my scene. Or have you been to Kunde Estate Winery in Sonoma? That's a beautiful place," Scott said.

Scott's home is spacious and modern, a multistory maze with few hallways but many doors and several flights of stairs. The stove in his kitchen sits in a counter facing the living room as if to a studio audience. He'd set up for us before we came. Bowls full of berries and crushed garlic sat next to empty appetizer plates, and a bed of fresh, garden-cut rosemary lay green and redolent on the counter. A half-finished bottle of old-vine California Zinfandel towered benignly over the scene. Well-used skillets waited empty and impatient on the gas range. We were here to learn how to prepare duck and, sadly, how not to.

Scott pulled the cleaned, Valley-shot duck from his fridge and began to carve it. His knife was handmade in Japan, a santoku, which means "three purpose." Made of laminated steel, it is the knife of a lifetime, and seemed heavy with its own excellence. The knife was difficult to look away from, and difficult to study. While he sliced red pieces of meat from the duck, Scott philosophized for us in an easy, hypnotic voice.

PLATE 4.13

"Most of the problems with the preparation of game stem from improperly prepared and handled meat. People think that to remove the game taste they need to cook the game out of the meat, but in fact, they are cooking the game into it. I try to show people that properly prepared game is as good as if not better than farm-raised meat. Another problem is that people don't know how to use the whole animal. They take a duck and just cut out the breasts. Well, the legs can be easily turned into a delicious dish, and the rest of the bird can be used to make stock for soup. It's just so simple."

The legs waited on the cutting board while the breasts were salted, peppered, and then tossed into the skillet. Duck, pepper, and olive oil scents blossomed in the air. Five minutes later, the breasts were ready and it was time for the delightful showmanship that separates the

the fireworks. But Scott's interest in converting us was misplaced. We already knew something of the pleasures of wild meat. The duck was devoured and I sucked the juice off my fingers.

"Now," Scott continued, "this is what an overcooked duck breast tastes like. You wouldn't even know it was the same bird."

Scooter and I looked on aghast as Scott mercilessly fried the innocent breasts. I did not need this lesson. But again, the story must be told. You, dear reader, anonymous heathen, child of the fallen world, must know: wild meat that has been overcooked tastes bad. Scooter and I were witnesses and reluctant prophets of this message. After three minutes I wanted to cry out, "Stop!" My fingers itched to pluck the meat out of the skillet. But no. The apocalyptic vision continued. The meat began to pull into

---

We were here to learn how to prepare duck and, sadly, how not to.

---

professional chef from the rest of us. Berries, rosemary, old-vine red, and a thick balsamic vinegar flew into the pan and were tossed about with the tender meat. After but a couple minutes more, the dish was done.

"Good balsamic vinegar is the key to a lot of recipes. There are two ways of getting it. You can buy the expensive, aged vinegar, which I never do, or you can simply reduce a decent, cheap vinegar. The taste is the same."

Scott arranged the food in an artful heap and I attempted not to faint as the photographs were taken. The drive had been long and we had skipped breakfast. My hunger made me resent such ceremony. I know what I want. Don't dress up for me. Just take me to bed and leave me exhausted.

This was a proselytizing dish. The breast meat was the occasion, and the berries, wine, and balsamic were

itself, to shrink. I could no longer look. I turned my head. The sound of the flesh cooking seemed like a lament. I wanted to stop up my ears, reverse time, whisper the secret into every ear on Earth, but I was powerless. Finally, Scott removed the overcooked meat from its torment and slapped it on a plate.

We reached brokenly for our forks and set the breasts on our tongues. They were chewy. Gamey. The moist, cherubic flavor we had cherished in the first dish had vanished. But I was famished. And I loathe waste. So I gave this ill-used duck a proper burial and the plates were empty once more.

Scott got right to work with the legs, which, unlike the gentle breast meat, benefit from long, slow cooking.

"Now, after the legs are browned, you want to bake

them for about forty minutes, but to save time, I went ahead and prepared some before you came."

And just like on television, a beautifully cooked pan full of duck legs appeared from out of his oven. He arranged these on a pile of rosemary and we contended with his boy for the lion's share of the scrumptious, melting meat.

"What's the strangest place you've ever cooked?" I asked Scott, expecting a story about Chile or Alaska or some far-flung corner of the earth.

"East Kentucky," he said dryly. It was a long drive into the deep woods. The restaurant was attached to a gas station. He showed up in chef's dress with a carload of strange ingredients, like crimini mushrooms. He was greeted gruffly by a trio of large women. There was whispering and potential mutiny. Finally, at the crucial moment, Scott convinced them that even though he was a man who lived in California, wore funny clothes, and liked to cook, he wasn't, in fact, queer.

"I grew up in Virginia and learned to hunt in the Blue Ridge Mountains. So when I'm in the South, I defend California, but when I'm in California, I defend the South."

Alongside his house, Scott keeps a terraced herb garden. Rosemary, several varieties of mint, thyme, marjoram, and basil. We left him there in a moment of domestic joy, with his dogs, looking up at his son, who had monkeyed up a tree and looked back.

He reminded me of a novelist in his later period. A tour through his cookbook confirmed that suspicion. His style has become more economical. Where his old recipes might have fourteen ingredients, he now makes them with five or six. The inclusion of wild animals into the world of recipes makes beautiful prose.

Soak quail in sherry, turning often.

Spread a thin layer of hoisin sauce over each pancake and roll with goose filling.

Ladle rabbit and sauce equally over each and top with goat cheese and cooked bacon.

Unless the nation's sentiments about hunting change radically, you'll never see Scott Leysath on the Food Network, despite his charm, expertise, and television manners. The irate calls and letters from a loud anti-hunting population are threat enough to keep the networks nervous. We passed Folsom Prison as we drove back towards Sacramento, tucked away behind rolling hillocks dotted with valley oaks. The prison used to graze its own cattle and raise its own produce before bone shivs became so frequent and trucked-in food so cheap that the prison administrators gave up autonomy. As we drove on, we pondered the things most Americans simply prefer not to see.

PLATE 4.14

PLATE 4.15

Aisu Pops

IN THE BEGINNING, THERE WAS THE TRICYCLE. After its near-miraculous construction in one of the Great Lakes states, a pedal-propelled portable freezer sat unused until Jaymes discovered it for sale on the Internet. The hour was late. The impulse was strong. And shortly after, the treat-trike arrived in pieces in Davis, California, and was reassembled. But what to keep from rot therein? And where to take those things? The answer was, if you haven't already guessed, frozen pops made with expired fruit and vegetables from the Davis farmers' market and sold in the same place. Frozen *vegetable* pops.

"Vegetables belong in desserts," claims Jaymes. "They just don't know it yet."

We were to meet Jaymes in Tucos Wine Bar and Café in downtown Davis, where Pru allows her off-hour use of his kitchen as well as the twenty-below freezers essential to pop storage. Jaymes designed her own molding process, which she credits for the purity of line and color in her brand, Aisu Pops, and the particulars are a closely guarded secret. The process takes twelve hours, which means that at eleven every morning and eleven every night, Jaymes is hard at work with knives, blenders, and a box full of squishy produce, making small-batch pops for her adoring public.

We were well into our third glass of wine each when Jaymes arrived, a small young woman whose hair was gelled into a crest high above her face and who wore a t-shirt sporting some undoubtedly hilarious cartoon in one of the Asian languages. Jaymes, we realized then, was cool. Much, much cooler than we were.

We introduced ourselves and Jaymes got right down to business.

"You want to try some pops?" she asked.

"Hell yes," we answered in unison.

She led us back to the storage room where the freezers were kept. She'd been saving some pops for our visit. Usually, her batches are so small and business so good that the pops don't age past their optimal freshness, but she was worried about a few that she had held back for us to try.

She held out an orange pop, stick first, to each of us and we unsheathed the treats from their plastic wrappers. The bright orange, tongue-shaped pop was strewn with small, dark specks. To my red-wine-parched mouth, the cold and wet of the pop was a sobering elixir. The ripe melon flavor was unmistakable, but there were other elements at work...was that cardamom and ginger I tasted?

"This is Heirloom Melon Chai," Jaymes explained. "I was just calling it Melon Chai, but after I added 'heirloom' it started selling a lot better. Here, try the Tangerine Beet."

PLATE 4.16

Plate 4.17

We went through half a dozen different pops, tasting flavors that ranged from basil, rosemary, and mint to mango, carrot, strawberry, and avocado. Her Lime Kefir Avocado pop was my favorite.

"The avocado pop is the hardest sell," Jaymes said, "but it has a die-hard following."

"These pops are exactly what I love about the Valley," I began. "The very thing that everyone on the coast hates about the Valley—the heat of the summers—is exactly why these delicious pops were created out here. The extremity of the climate produces a culture of relief—and what is

halfway out of the passenger-side window, could take her photograph with a minimum of blurring.

Jaymes has developed great relationships with many of the farmers at the market. Not only do they supply her with most of the raw materials for her pops, but they are also some of her most devoted and daring customers. We followed along as Jaymes delivered a Strawberry Basil pop to Annie from Good Humus Farm. Jaymes handed off the pop and then began sampling Annie's wares.

"Why are your strawberries so good?" Jaymes asked.

"To drive you wild!" Annie replied between red,

Jaymes is hard at work with knives, blenders, and a box full of squishy produce.

---

better in one-hundred-degree heat than a Strawberry Basil pop? In the continual fog of San Francisco, no one would think of gourmet fruit and vegetable pops."

Jaymes was unimpressed with my rhetorical flight. Her vision far exceeded mine.

"I'm actually trying to change that. I think pops should be enjoyed all year and in every kind of weather."

We left Tucos sometime after one in the morning, our lips and tongues stained Tangerine Beet red, and new flavor combinations on our minds.

"Habañero Lime."

"Winter Squash and Cardamom."

"Cucumber and Mint," we bandied back and forth as we searched for a place to rest our heads.

WE MET JAYMES AGAIN on a Thursday as she was preparing to attend that evening's farmers' market. I followed her and her tricycle down the sleepy Davis streets in the Jetta, matching her speed when I could so that Scott, stuck

sticky mouthfuls.

Before we left, I asked Jaymes what was next.

"I don't know," she said. "Maybe a gourmet corn dog?"

JAYMES IS A GUIDING LIGHT for the future of the Valley, and for the future of food, and our edibilism, if we are lucky. The closed circuit of farmers' market to artisanal food to farmers' market, animated by the precise inventions of a gifted creator, has the potential to form new traditions, new expressions of the polyphonic hyperculture that characterizes the Valley and most of California. Whereas John Pereira and Michele are the activists of tradition, and Pru is a self-aware convert to fresh and local cuisine, Jaymes seems to have grown up in a time when the revolutionary ideas of the back-to-the-land, organic, local, and seasonal thinkers and makers and growers have been internalized, taken for granted as valid and true: allowing, perhaps, a greater freedom for invention. Or maybe she is just the agent of destiny for an oversize tricycle from the Midwest.

PLATE 4.18

Plate 5.1

FIVE: Found on the Road

Plate 5.2

Prayer for Entering Highway 99

Please Lord,
Let not death be
A tomato truck.

Volta, Unincorporated

I WAS MAKING A FRIGHTENED RABBIT'S RUN out of the Valley, rattling down small roads through the desolate miles, westward, away. My first exploration of the lower San Joaquin Valley, the arable lands of Fresno and Tulare Counties, had ended in disaster. I'd thought the Sacramento Valley had shown me horrors sufficient to my task, that I was now inured to the spiritual torpor induced by miles and miles of monoculture. I had witnessed the mass melon grave. I'd seen pickers resting under the machinery, the only shade for miles. I thought I was ready.

But in the citrus groves extending from Lindsay north to Exeter, and then on every side of Reedley, Orosi, Sanger, and Selma, I encountered a new terror. Every orange on every tree spoke to me of its destination, and I traveled with them through time and space. I was driving down pale, dry orchard roads, but I was also an invisible attendant at every breakfast table across the country. The cataract groan of millions of glasses filling with juice, the shriek of every spiral-torn peel, the thunderous applause of teeth against pulp—the hive sound of that simultaneity increased with the hailstone clatter of unpicked oranges falling against the hard, lifeless ground. In searching for the edges of bounty, I had stumbled unwittingly into bounty's core. Flee! I pointed the car towards home and threw up clouds of dust to obscure my retreat. Mercifully, the citrus groves eventually gave up the chase, but instead of their terrible globular monotony, I had now driven into a dust-riven, hard plain of emptiness, a peach pit pressed flat. The only break in the agriculture of the apocalypse was the smell of molasses and manure from innumerable dairies of Holstein milk and misery. Land o' lakes my sweaty ass.

The towns along Colorado Avenue remain a blur—San Joaquin, Tranquility, Helm, Ingle. Signs began to appear shortly after for Mendota, the Cantaloupe Center of the World. What I needed was a thriving little downtown with a dark bar staffed by a pretty woman with a community college education, a kid or two, and a piece-of-shit ex-husband in Madera, but the sewage disposal centers along the Fresno Slough augured poorly. Mendota may be a nice town, but I saw only the paralytic immobility of its streets, the invalid stare of boarded-up shops and the usual broken-window ruin that agribusiness brings to its devotee cities, so I sped past towards Firebaugh. No hope there either. I took Avenue 7½ towards the Buttonwillow Drain on a whim, crossed it, despaired, and backtracked to Highway 33 North, a broken man.

I turned off 33 onto Volta Road to avoid Interstate 5 and crossed a road named Badger Flat. If I could make it to the Pastime Club in Gustine, I'd be able to make it home safe. Not that there was any escape. The Valley may fuel this ugly civilization, but the Cities of Enlightenment along the San Francisco Bay are the engines, and their self-righteous bubble life is only made possible by the heedless predation of rural places. *Broken*, I thought. *Every goddamned thing under the sun is broken.*

a marker of class. These old Chevrolets and Buicks still ran, these old houses still sheltered, and a well-made thing, be it wall or saw, should last a century or more. Agriculture itself is something this civilization wants to set behind it. This deletion is a clever way to ignore entire regions and peoples, to place them outside of time itself. But what an impoverishment, to find currency only in the infant tantrums of new technologies and fashions.

The boys rode by again, taking no heed of me, shouting to one another in English and Spanish like a bilingual version of my own small-town years, which are already considered antiquated and irrelevant though they were only two decades ago. In another two decades, these boys will still be riding around this town, or others around another much like it, and

---

Animals are a great benefit to the poor, and by the poor I mean mankind.

No bigger than a flea on the enormous dachshund of the Valley—only four small blocks by three—with roads of an ancient pavement covered over with sand, Volta, Unincorporated, arrested my deranged flight effortlessly. *This place*, I thought, *breathes its own air.* A gang of little boys rode bicycles lazily down the empty streets, steering one-handed around potholes and chickens. The setting sun made everything bigger, and slower to leave the senses. In the yards of houses variously upkept, there grazed goats and an occasional head of cattle or two. Nobody's chickens ranged everywhere and roosted in old cars resting on cinder blocks. Cats with speakeasy attitudes strutted from slanting light to long shadow.

Time, it is said, has stopped in a place like this. The sentiment had always bothered me, but I hadn't, until that lazy dusk in Volta, thought it through. Time, it seems, rather than being measured in hours, days, or years, has come to be measured by material goods, the implication being that contemporaneity, and therefore relevance, is a luxury item,

much different, in the year 2027, 2057, 2120, 2182...Volta is the future, or one of them, anyway.

I TOOK SCOTT TO VOLTA a few months later, a little embarrassed by my desert enthusiasm for what even I knew to be a modest, muddy spring, but I should have had more faith in my comrade. Too often I am content to be Emerson's transparent eyeball, but Scott reminds me that we should always at least *try* talking to other people.

Inside the gutted husk of a feed store, Rosie runs a bare-bones convenience store, offering a few candy bars, some chips, a couple six packs of beer, some soda, and cigarettes.

"Oh no," Rosie said when we walked out of the sun, "You're not taking my picture."

"Are you sure? You haven't gotten to know me yet." Scott rejoined.

Rosie, an old Portuguese widow, has lived in Volta since

three days before forever. She lived with her husband up north when he was with the Corps building one of those big dams, but they came back. She's the Chamber of Commerce, mayor, Neighborhood Watch, town historian, and gossip of Volta, and she controls the water too.

Rosie on the town before it lost its railroad stop, post office, and incorporation: "Not too different."

Rosie on the town drug problem: "Some people think that just because we don't have any police they can do whatever they want, but I see it all and I call. Those out-of-town cars come in the middle of the night and I write down their plates and call them up. If I think one of the residents are involved, I'll shut off their water. I don't care."

Rosie on photography: "Oh, I love pictures. Ever since my husband died, I've been working on my wall. Every night I put all our photos up on the wall in a new order. I don't sleep much anymore, so I just spend all night with those pictures, trying to get them right."

Rosie on the chickens: "Those chickens have always been here, but I don't like them around my place. One of the women here feeds them—claims they ain't hers, but feeds them anyway. But sometimes one of those chickens just disappears, and she comes around asking about it. I tell her I don't know, but we all know. Sometimes you can even smell one of those chickens cooking."

PLATE 5.3

AFTER TALKING TO ROSIE, I WONDERED how Volta could seem not only so contemporary, but even on the cutting edge. The animals...the humbleness of its size...were Volta incorporated, laws would prevent the residents from keeping a goat or two in their backyard, the physical distance from non-luxury animals being a subconscious condition of today's civilization. But animals are a great benefit to the poor, and by the poor I mean mankind, for it is only in the necessity for food, clothing, company, and vocation commonly called poverty that we are known to one another. Should the oil run out, should this vainglorious society wreck upon that iceberg, we'll bring the animals back into our lives. For their meat, their milk, their fur, wool, and hides, for their bones, for their warmth, company, and for the unique, unspeakable knowledge of ourselves found in their eyes. They provide for our every necessity, and we owe to them the greater portion of what we call human. It is a comfort to me, the odds that they will return to us, though we scarce deserve them.

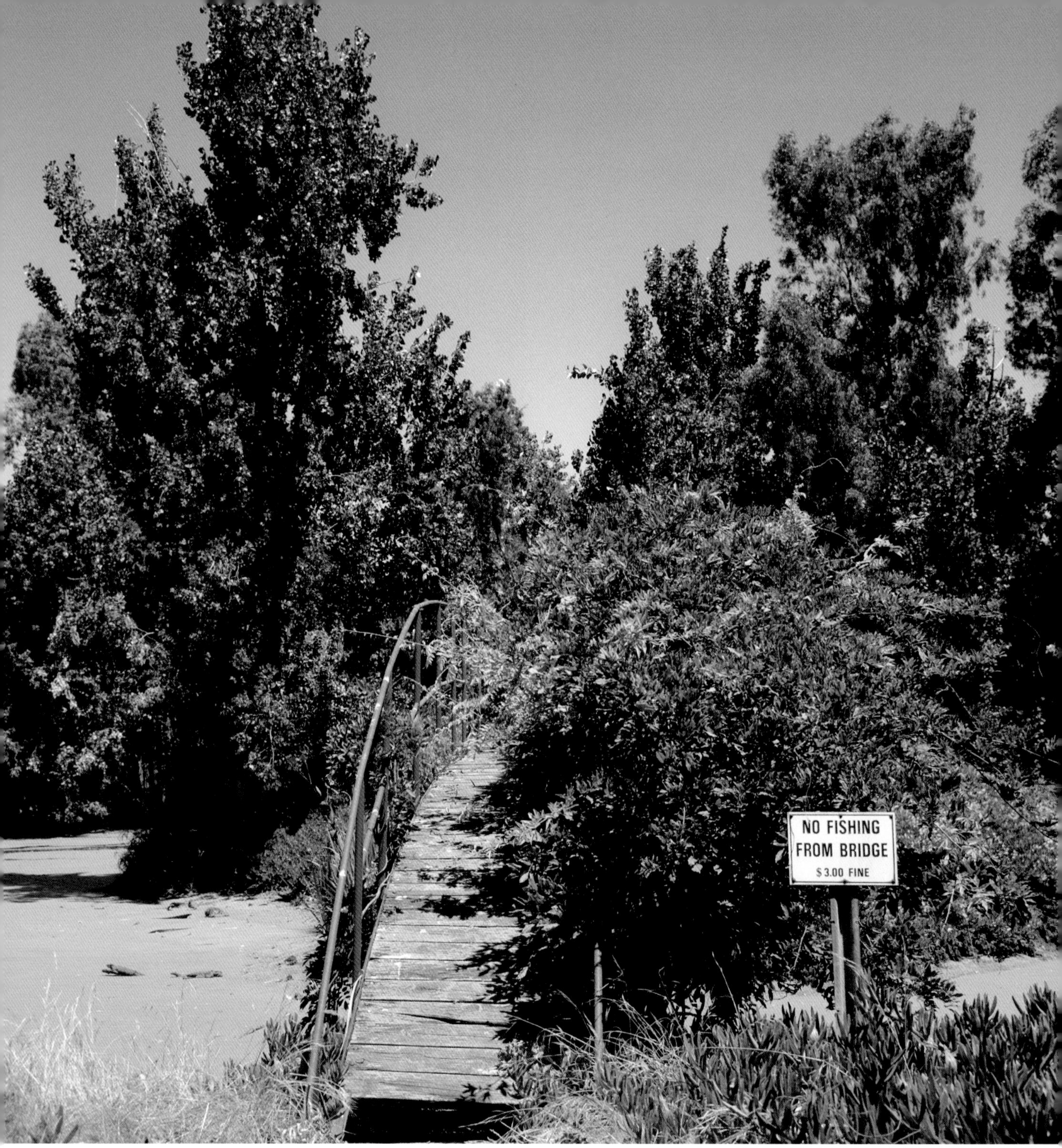
NO FISHING
FROM BRIDGE
$ 3.00 FINE

PLATE 54

The Egret Palace

The Egret Palace rests behind a high wall of eucalyptus trees along a quiet bend of Inland Road. It is known to winged beasts far and wide, but it is the egrets, those supple, knife-like specters of irrigation ditches, that lend it majesty. To know the destination of birds, and to witness their formless coming and going, is to realize the awkward strangeness of our reasoned network of roads. Watching the airy cartography of the egrets makes one want to abandon automobiles and begin walking again. To give up on straight lines. We were not looking for this place, nor hoping for these insights. We arrived at the Egret Palace unwittingly.

Two ghastly figures stood threateningly over the low mess of a tomato field, dressed in long white coats, their arms cast out from their chests in a gesture of intimidation. I stopped the car to get a better look. Scarecrows, we realized, but that did not lessen the eerie mystery. We'd never seen a single scarecrow in any Valley field, and certainly not one dressed in long fluttering robes. These stood perhaps fifty feet apart and each faced the road. The tall lonely crosses and the burning white gowns hanged upon them evoked the Deep South in dark times, the costume of the Klan and its image of terror brought together in a strange marriage. Certainly, I was scared, but not so scared that I wouldn't steal a ripe tomato if I had the opportunity, means, and motive. But what birds threatened this field? Why here and nowhere else?

Scott got out of the car and waded into the tomatoes with his camera. I lounged in the shade with my notebook and pondered the scene. It took me nearly a minute to discern the wave of cacophonous jungle sounds coming from the trees behind me, to awaken to the world around me. I walked down the road to be nearer the sound, my steps instinctively careful and quiet. I tucked my head into a gap in the solid windbreak. Tall eucalyptus trees were widely spaced throughout the clearing, providing an almost unbroken canopy of thin leaves. The noise increased. I took one step into the glen and set off a tornado of small, white dragons. Six dozen egrets resettled at the very top tufts of the trees and pretended not to look at me. I retreated.

I was a little giddy when Scott returned, squatted by the open trunk of my car, and began to change his film.

"You ready for some fast shooting?" I asked him.

We approached the clearing. The birds were less nervous now. The whirlwind I had seen failed to form, but the sky was filled with crisscrossing streaks of white. Further into the grounds, the earth was white and powdery from the effluent excretions of the avian court. We noticed the litter of eggshells among the shit-caked eucalyptus leaves and the general cacophony became more targeted and personal. Egrets flew towards us, their wings slicing the air near our heads. We looked more closely at the trees and saw a host of rough nests in low branches. Clumsy half-fledged egrets ran from us.

"This is a special place," Scott said. "We shouldn't be here."

"I think there is a body of water further on," I said. "Maybe we can skirt around to the left."

Along the left side of the citadel an alfalfa field endured the gossamer attentions of a butterfly horde. They flew like pieces of burnt paper. Half were the color of eggshell, the other half yellow as an egg yolk. A deeply rutted farm road wound around the trees. Firewood and brush were piled separately alongside the road at regular intervals. We found another gap in the ring of eucalyptus and climbed a brief acclivity. A lagoon covered in green moss and crowded with trees waited there for us. We crept on our hands and knees closer to the water. A quick-eyed pair of ducks spotted us and flew low and fat across our vision, resettling with some

With trepidation we set wary foot on the soft wood of the thin bridge. Our hands gripped the iron railing. We crouched to avoid the overgrown vine. We began to cross, knowing we faced almost certain ickiness. Halfway to the island a plank finished its slow rot and plummeted into the water beneath my foot. We continued, and soon we stood on the island, a strange hillock completely covered with Indian paintbrush. This was the best view of the lagoon, but it added a finality to our observations. Our presence on the island made it clear that it was time to leave. The enchantment was broken. We recrossed the bridge, took one final look around, and walked among the butterflies along the farm road back to the car. The clinicians, as I'd begun to think of the scarecrows, burned just as whitely in the tomato field as when we first saw them. They must

Two ghastly figures stood threateningly over the low mess of a tomato field.

noise not far away. A downy feather fell onto the water and the wind helped it walk across the moss towards us. It crossed the entire pond in this manner. The branches of the trees were all stained white and shook with the restlessness of the egrets. The birds relieved themselves into the water with loud plops. I don't know how long we sat there. What are watches to wilderness?

We left our perch and continued to follow the ring of trees. Around the bend the familiar site of a working farmyard jarred strangely with the nearby jungle. Prefabricated tin sheds and large tractors sat outside a barely visible single-story house. A huge metal pipe crawled out of the water and into the yard. A small dock became visible along the far edge of the pond, and the path we followed led to a wooden bridge overgrown with some kind of flowering vine. The bridge led to a small island. Posted at the foot of the bridge was a sign.

NO FISHING FROM BRIDGE

$3.00 FINE

be effective, I thought, for there were hundreds of birds across the road, but not a single egret in the tomato field. The bird sounds lessened as we approached the car but did not go away.

The entirety of the Egret Palace rests on perhaps half an acre. Some strange farmer built and maintains it. Architect, groundskeeper, artist. Did he plant the fast-growing eucalyptus as well? There is no question of ownership there. It is a gift to the birds, and they are shaping and using it according to their needs. Does he sit on his wooden dock and stare up at the woven nests of the egrets? Is he heartened then, that he is one of them, an animal, glad in the company of his fellows, noisy, restless, and easily frightened? There is nothing edible here. No ridiculous fantasies of the farmer trapping the birds, force-feeding them, then smoking them in secondhand wine barrels. It was just a beautiful thing that happened to us, and we left hungry.

PLATE 5.5

Plate 5.6

Let My People Go

We walked into the Cigar Box, covered in white dust and streaked with sweat. Our skin mimicked the jigsaw pattern of the parched earth all around us. It was October in Yuba County and the earth had forgotten even the face of rain. In the meantime, it had been tilled by iron and steel, trampled by countless feet, pulverized by the wheels of trucks and tractors, sprayed with pesticides, herbicides, and nitrogen, and eroded by endless winds. So, besieged in its own territory, the dirt splintered and took to the air, blackening the phlegm of the inhabitants and draping sunrise and set in the red of war. The valley was one vast crematorium and smelled of fried onions...so we grabbed a couple stools at the bar and ordered a line of beers and some ice water and keep that ice water coming.

The barkeep was telling a customer that he'd seen Charlie the night before, fresh out of his three-month stint in County. They seemed to agree that whatever Charlie did had had to be done, and the fact that he'd served any time at all was a perversion of justice. Scott interrupted them to ask about a rumor we had come across earlier in the day.

"You guys ever heard about anyone harvesting crayfish when they drain the rice fields?"

"Heard about it? Hell, I've done it myself. When they drain those fields, the crawdads swarm outta there and go looking for more water. You just get a snow shovel and scoop them into a trash bin. A lot of Asians are out there. They'll bring pots and boil them right there on the side of the road. The crawdads are big, too. Eight, ten inches long."

Scott straightened on his stool. Despite the layers of dirt, dust, exhaustion, sunburn, and heatstroke, his spirit remained strong.

"Do you know where any of that action might be taking place?"

"Nah," the barkeep replied, "I haven't done it since I was a kid. I think all the rice fields around here are drained already anyway. Might be a few farther north."

We spent another couple hours in the bar, emptied a few more bottles, consulted our maps, washed our hands, arms, and faces, and talked about this new lead. Crawdads. Rice fields. Snow shovels. Families cooking on the roadside.

"We've got to find this," Scott said. "This is the keystone of our whole trip." Scott's blood was up. He was playing nervously with his camera and eying the biker with a swastika tattoo at the pool table. We left the Cigar Box and squinted into the deathless Valley sun, restless as thieves awaiting the next big score.

We arrived in Knights Landing ravenous. The taco truck parked in the empty lot of a defunct auto shop lured us in with a siren scent of carnitas, carne asada, and barbacoa tacos, and we circled our prey like wolves. A couple plates of cornmeal, beef, pork, salsa, and some lime soda later, we looked around with human intelligence once more. Across the bending highway stood a dilapidated building that resembled an enormous shed. We could see a sign advertising live crayfish on one door. The gate to the high fence that surrounded that section of the building was padlocked, but a few windows were exposed along other walls. Through the dark, warped grime of ancient glass we could detect a

the building contemplates theology. Cremation, burial, annihilation, reincarnation? Spiders lawyer its demise. The windows swell and muddy with age. The bait shop slowly acquiesces to its architecture a mile away from the river that runs through Knights Landing, California."

———

A crawdad is a cipher for human sympathies. The enormity of his claws prefigures the beseeching hands of Rodin's Burghers, while he himself is as small as a doll. The crawdad is the peasant of river, creek, and Delta, always bowing, a permanent supplicant. His eyes are bright with the desperate cunning of the scavenger

———

"You guys ever heard about anyone harvesting crayfish
when they drain the rice fields?"

———

pool table, a room with tables bristling with the legs of overturned chairs, and a long bar. A fallen sign read: BIKERS WELCOME. This bait-shop turned highway-and-river bar had obviously been closed for some time, and I wondered what this modest crypt of good times meant to its merchandise. The crawfish. The crawdad. The crayfish. I wrote:"The crayfish rejoice. The gaoler, the wolf of the rivers, the tyrant has fallen! They raise their claws above the water and clap them together—commemorating each anniversary of the bankruptcy, death, or act of crayfish god that padlocked the doors of this fisherman's hole. But not all. Members of a small society of crayfish paint themselves darker with the river mud, believing that if the ogre has no use for them, he is likely to—unthinkingly, clumsily, with poison—annihilate them in passing. They are organized. They have a newsletter and hold fund-raisers. Meanwhile,

cowering, furtive, in mud or crevice from his scaled predators, but striking with murderous precision from these selfsame alleys when weaker creatures idle by. He wallows gluttonously in orange, edible beds of salmon roe. Crawdads cram themselves by the hundreds into fishermen's traps, drawn by individual avarice into collective indenture. Served in a plated heap or steel bucket, how like a red vision of man debased. Our instinctive loathing of the homunculus fires our semi-cannibalistic zeal. We tear the crawdads in half, prize the tender meat from their tails and devour it, and then we suck out the guts and shatter the claws. The melted butter speaks of luxury, of pillows and silk, the hot sauce to the fires of lust, pain, and fear that drive us on to the next crawdad, and the next, until finally there is naught but shards of exoskeleton—the masks of comedy and tragedy mingled now inexorably, the mirror broken.

———

PLATE 5.7

In 1986 a diminishing Isleton stood at the crossroads of extinction and made a pact with the devil. This Delta town of under three hundred souls would not go quiet into that long night. The incantations were simple. A little hype. Some t-shirts. A little insta-myth and an invented history to supplant the flood survivals, Chinatown, and days of luxury past. Six months later, on Father's Day weekend (and oh, what a father!), from the anguished womb of Isleton issued the Annual Crawdad Festival. Twenty years later, the small town festival resembles Gargantua on a three-day bender. Hundreds of thousands of people make a wasp's nest of the streets.

"We, uh, ran outta crawdads last night. We don't get any more until Monday," apologized the gentleman behind the bar. "But we might have some more tail meat for the crawdad melt."

Scott took the man up on his offer and I ordered God knows what else with an appetizer of breaded and deep-fried Lodi asparagus, side of ranch. Scott took a bite of his toasted, fishy, cheese-and-crayfish-tail sandwich and then taunted me with his mouth full.

"Oh, you so should have ordered the melt."

As we passed Bob's "The Master Baiter" Bait Shop, Scott paused with his camera near a window. Supplicant

---

The crowds heave and sag and wet the streets with domestic lager,
only to pave them over with the red shells of discarded crustaceans.

Local crawfishermen supply historic restaurants like Ernie's as best they are able, but the hordes of vendors and festival fly-by-nights ship theirs in from Louisiana, an irony exploited by dyspeptic big-city journalists in their yearly coverage of the event. The town is cordoned off. The crowds heave and sag and wet the streets with domestic lager, only to pave them over with the red shells of discarded crustaceans. Most of the locals flee. To make the simulacrum complete, women young and less young bare their breasts for beads and Cajun music fills the air. The festival is the reanimated corpse of Mardi Gras past.

Or so I have reconstructed from archival materials, interviews, and survivor accounts. We had no intention of elbowing our way through the breast-baring beaded beery buttery hordes during the festival, but Isleton Joe's serves the crawlies up year-round. We bellied up to the bar and bellowed our order to any and all that would hear.

no longer, a crawfish had raised his pinchers high and proud against the wall of his prison. Scott began in a low baritone to sing: *Let my people go...*

The rest of the story is a sadness. The crawfishermen wouldn't talk to us. Gigi West, one of the founders of the festival, former owner of Ernie's, and creator of the crawdad melt, wanted to prepare us a feast, but first the spring rains washed out the traps, then the festival exhausted her, and finally the truck with all her pots broke down in Monterey County. A shiny rice salesman at Tucos in Davis promised us the world at two a.m. But we could see in the bored eyes of the young woman with him the worth of his words.

In the last days of our journey we returned to Marysville. Behind a Chinese temple a tall dike has been erected to protect downtown from the Feather River. A wooden staircase leads to the top of the dike. Atop the dike runs a railroad track, and on the other side a star-thistle-choked floodplain eventually gives way to the river itself. Three bridges are visible from a small, pebbled beach there. On the Yuba City side of the river someone had sculpted a dirt-bike course. The bikes were making an elliptical poetry of flight and fall as Scott and I stood on the opposite beach contemplating the trucks, cars, and trains crossing distant bridges. We contested each other at skipping stones across the afternoon surface of the slow waters, Scott handily the winner. That bright, perfect scene that had appeared so inevitable to us in the Cigar Box a year earlier took on a sepia hue and then faded to white altogether. We made peace with it then, the dying of

> Oh, the grapes were sour indeed.

so bright a fire, and threw our resentment along with our rocks into the river. It had been, perhaps, ordained thus. Our vision had been one of bounteous imperfection, and we had reaped such a harvest that the longed-for scene—crawdad hordes, snow shovels, families cooking on the side of the road—seemed stiff and contrived. Oh, the grapes were sour indeed, but sweetened, slyly, by the unspoken knowledge that autumn gave way to autumn. We would be here again, we didn't say, we would be here again.

PLATE 6.1

SIX: Killing, the Beginning, and the End

PLATE 6.2

Chico Lockers and Sausage

I HAVE DANCED MUCH AROUND THE MOMENT between an animal's life and its death. We visited Don Lopez, whose land is devoted to the raising of sheep and goats, and the celebration of their forms. We saw the Escalon market, a gateway to that critical moment, and talked to Debbe, but we had yet to witness the slaughter ourselves. Skipping a step, we'd eaten Basque Lamb and sopas. Now, moving backwards, we were at the step where the corpse becomes cuts of meat: the butcher's.

Deer season is not the time to beg butchers for an interview, but in the hierarchy of edibilism, a hunted animal processed for personal use is a finer thing than marinated chicken and links of sausage from certified slaughterhouse meat. Many ranchers also use butchers to process their own animals. Often a butcher and a slaughterer will travel to the ranch and kill and then process the meat of an animal for a fee. The meat stays with the family. It is illegal to sell the meat of an animal killed this way and to sell meat from hunted animals is also against the law. There have long been laws governing how animals may be killed and what meat might be eaten, but in the United States meat is consecrated by an industrial stamp rather than by religious rite. The consequences of regulation are not limited to the real or illusory protection of public health, but also include the ostracizing of those hunters and ranchers who feed their families outside of the industrial system.

We had sought to witness a ranch slaughterer at work, to be present when a living, breathing, thinking animal fell dull-eyed and heavy to the ground, but the animal rights movement has rescinded the trust of strangers. I had been holding myself tense for that moment from the beginning of our journey, half an eye always on it, and when it became clear that we had neither the luck nor opportunity to participate, I was more than a little relieved. But if there are those among us who, while refusing to see all meat as murder, wish to reclaim meat from the industrial slaughterhouse, the decriminalization of food from animals killed at home, either pasture or wood, one at a time, would be a fine place to start.

"Goddamn it," I said to Scott after Bob Dee left us with a sample of gourmet German bologna. "I was happy thinking that I hated baloney. Now I like it."

"Yep," Scott said. "This is the best bologna I've ever had. Seems like we're running into a lot of personal bests on these trips."

Bob Dee is the owner of Dee's Meats, in Galt, a little town southeast of Sacramento. He specializes in European-style sausage but sells a variety of products from jerky to all-beef hot dogs and runs two small butcher shops with the help of his son. He gladly took us around his shop, showed us smokers, extruders, vacuum sealers, and grinders, and catered the tour by cutting off slices of whatever was available for us to taste. A ranch-slaughtered young pig hung from a hook

> The body of a deer is a plump feast, conceived to be brought down and eaten.

in his cooler with a big red stamp on its head that said NOT FOR SALE. In a separate room, two halves of a deer carcass hung side by side on a motorized track.

"I don't take many deer these days," Bob said. "I'm getting too old for it. But I've got a few people who have been bringing me deer for thirty years. I can't really turn them away. I'm starting to look forward to getting out of the business altogether. I don't know how those slaughterers can do it. By the time I see the animals, they're just meat. But I don't like seeing them killed. I had a friend who was a slaughterer, and he said he used to dream every night of murdering human beings. Every night. He retired a few years ago and I asked him if he still had those dreams. After he quit, the dreams stopped."

THE NEXT BUTCHER SHOP we tried was Tom Ball Custom Meat Cutting, in Orland, on the way back from our futile salmon expedition in Red Bluff. It was a modest operation, with only a couple display freezers selling some hamburger, steaks, and roasts, but the meat locker was visible behind a glass wall, stuffed wall-to-wall with deer. All of our hustle and charm got us nowhere.

"You should talk to Dave Dewey in Chico. He loves the limelight."

So we did.

WE PULLED INTO CHICO Lockers and Sausage and were not surprised to find Dave chatting up some hunters in the deli when we arrived.

"Give me a few minutes, and then I'll show you around."

As those few minutes gave us a chance to order a sandwich from his deli, we gladly took the opportunity. When we were halfway through our meal, Dave returned.

"Well, what's taking you so long?" he asked.

Dave's operation was large and thriving.

"There used to be three butchers in every town in the Valley, but in the last thirty years most of those have closed. We have all the business we can handle," Dave said.

When Dave showed us his meat locker, filled with deer and swine, Scott made soft mewing sounds and asked to be left alone for a while with his dream come true. In the meantime, I chatted up Dave, trying to assuage the suspicion I detected in his eyes, as he and his employees made snack sticks.

"What did you two say you were doing this for?" he asked. I explained again and he offered me a smoked snack stick.

"This is great. Do you have any of the spicy ones?" I asked as Scott reappeared, requested a stepladder, stuffed a snack stick in his mouth, and disappeared back into the locker.

"I thought you were some of those animal rights people for a second," Dave confessed, "but when I saw you devour those snack sticks, I knew you were all right."

"A deer's on its way," Dave's son said, running in from the deli.

"We've been hoping to see someone bringing in a deer to be processed," I said. "Do you think we could wait around?"

"Sure," Dave said.

LARRY WILLIAMS AND HIS WIFE had been hunting in Idaho and were bringing in two deer wrapped in white body bags. He backed his truck up to the loading platform of Chico Lockers and Sausage and wrestled the deer out of

PLATE 6.3

the truck. Dave popped out of a side door with a saw.

"Do you want the antlers?" he asked.

"Might as well," Larry answered.

Dave knelt next to one of the deer and set the saw above the antlers. The antlers and a section of the forehead slid away from the rest of the body like butter and Dave quickly did the same to the other deer. Larry and Dave discussed ratios of ground venison, steaks, sausage, and snack sticks until the optimal percentages were agreed upon.

The deer lay soft and still on the shaded concrete of the wet, and inquisitive. Larry traveled hundreds of miles to find and woo these two deer. He had watched their living steps, counted, perhaps, their living breaths, and then taken their lives. There, on the ground of their deaths, Larry prepared their bodies, holding each against his own, staining his clothes with their blood, and covered them with white bags, then carried them down a mountain to his truck. After Dave divests the deer of the last remnants of their bodies, their purple-red meat will feed Larry and his family for the better part of a year.

If there are those among us who...wish to reclaim meat from the industrial slaughterhouse, the decriminalization of food from animals killed at home...would be a fine place to start.

---

loading bay, their opened faces revealing bone and brain and half a milky eye. I contemplated their beauty. It was not obliterated by the revelation of their meat. It held. I imagined their legs folded awkwardly beneath them in an imitation of rest beneath the white body bags. The legs of a deer are too thin for his body. The body of a deer is a plump feast, conceived to be brought down and eaten. The neck is long and curves shapely to the sky, the upward sweep terminating at the tips of the coveted antlers, arrayed like lightning bolts in a summer storm. Two pair of these antlers sat abstracted on the edge of the loading bay. The eyes of a deer possess a dark gentleness that stills and dissipates discord. Beneath the eyes the nose lopes down a ways and then waits, soft,

Despite a knee-jerk grimace at the surface of gore, at my own fear of death, the slain deer, their detached antlers and the revealed soft red and white of blood, meat, bone, and brain, were not ugly. There was admiration and even casual and unexamined affection between Dave and Larry and the two bodies on the concrete dock. The transaction was even theatrical, in a way, with the dock a stage and the exchange, carried on so calmly, containing secret meaning. Actually, it was beautiful to watch, and the red of blood, where I stood far from its iron scent, strange viscocity, and tendency to turn brown at its edges, was simple, striking.

But I kept my distance and came no closer than I had to.

PLATE 64

Plate 6.5

Junior and Ladies Pheasant Hunt

DALE WHITMORE CRADLED A PHEASANT in his arms and addressed his volunteers at the Annual Junior and Ladies Pheasant Hunt.

"There are a number of ways to set a pheasant, but the Department of Fish and Game has decided that this one is the best. You take the pheasant and gently fold its head underneath one wing, and then hold it there for about thirty seconds."

The pheasant held tight against his chest, Dale continued his instruction in a softer voice.

"Always set the pheasant away from the truck. The easiest way is to start at the far end of the field and then slowly drive back the way you came, setting along the way. Now, let's see if I've done it right or if I embarrass myself."

The pheasant seemed limp and uncaring as Dale kicked up a small patch of grass with one foot and set it gently underneath the improvised blind. With one quick movement he opened his hands and took a quiet, stooped step backwards with a gesture that was half impresario, half delighted father. The trick had come off. The bird slept there under that thin patch of dry grass while Dale, consciously forgetful of its presence, assigned each volunteer to his section of the floodplain. The rest of the pheasants sat quietly in long wooden boxes on a trailer. The volunteers were to set ten pheasants in each section, and each section would only hold three hunters, so the odds of at least getting a shot at a bird were high for each hunter.

Trucks full of mothers, fathers, children, friends, and dogs were streaming in steadily while we rode out with Dan Silva, his son, his daughter, and their new dog to set some pheasants. Dan didn't use the prescribed method. He simply held the pheasant under a clump of grass for a minute and then walked away. The first pheasant wasn't fooled and exploded from the grass, flying high, loud, angry, and straight for the highest branches of a creekside cottonwood, but the rest stayed put and would have to take their chances.

Back at camp the participants ate donuts and drank coffee or soda while a pair of humorless and efficient women like those so often in charge of registration everywhere made sure everyone had their ducks in a row. Sections were assigned and maps were dispensed. Dogs filled every empty space. The crowd exuded pride, anticipation, caution, responsibility, and the unmistakable radiance of hunter's orange. Flimsy communal orange vests were provided for those, like us, who didn't come prepared. Mine fit like a bra while Scott's more resembled a string bikini top. Luck of the draw had given young Mr. Silva a section all to himself, and so his father, his little sister, a writer, a photographer, and an untrained bird dog pup all set out to watch him hunt. Such pressure would have daunted me, but he did not seem burdened.

The section was wild with thick golden grass, dry creeks, and wild grape. A road of mowed grass zigzagged uncertainly across the field. The dog accompanied the boy while the rest of us kept our distance. Dan was disappointed with his new dog. Their old dog, Misty, had been a legend in her own time and the new dog had a large collar to fill.

"There was this one time Misty had something cornered in our neighbor's bushes. I gave her the go-ahead and she dove right in there. There was screaming and howling and feathers were flying everywhere. Suddenly she slinks out and gives me this look, but I told her to get back in there

one shot, no hesitation. Unfortunately, there was no opportunity for a repeat performance. Though ten pheasants had supposedly been set, not one other was found in the entire field. We kept at it for almost an hour, but when we finally heard other groups of hunters start back for camp, we joined them.

The atmosphere at camp had devolved into pure pleasure. The business of the day was completed, the bodies of felled pheasants hung from every hunter's hand, and there were hot dogs and hamburgers coming off the barbecue. Once all the hunters returned there would be a raffle for various

---

The odds of at least getting one shot at a bird were high for each hunter.

---

and in she went. Eventually her quarry darted out. It was a big tom turkey half as tall as me. Misty was hot on his tail, though. She chased him through tennis courts and yards and bushes, but she finally got him."

"What happened to the turkey?"

"Oh, I can't remember," Dan said slyly.

"OoooDooooDoooDooooDO" the pheasant shouted before filling the air with the explosive flapping of his feathers. In one fluid movement, Dan's son brought his shotgun to his shoulder and pulled the trigger. The sound of the shot had not left our ears when we saw the pheasant plummet stiffly out of the air and disappear beneath the grass. The dog jumped excitedly but uselessly about the young hunter's feet as he calmly found the lifeless pheasant where it fell, posed for his photo, and then stuffed the bird into his game pocket.

"Did you see that?" Dan asked, hovering about three inches off the ground with pride. "Did you see that?"

It had been an awesome performance. One pheasant,

donated items.

Scott and I moved about the crowd, chatting up the hunters, admiring their bounty and playing with their dogs. Scott was moved by the hale and hearty day's work in the fields.

"I need to start hunting again," he said. "I don't think I could kill a deer anymore, but I wouldn't mind taking down a few birds."

One of the hunters who wasn't staying for the raffle gave me his ticket. I couldn't say no, but the gift did leave me in a predicament. I have no luck at all when I want to win something, but every time I expressly do not want to win, say, a three-by-five framed oil painting of wild turkeys in a wooded scene, I surely do. I didn't even have to double-check my ticket when they called my number. I simply walked up, claimed my prize, and tried to avoid the gaze of anyone who might envy my draw.

I had another problem. In order to really capture this moment in the Valley, Scott and I were hoping to secure a dinner invitation from one of the hunters. As they could not possibly guess our intentions, I had to find a way to invite

PLATE 6.6

PLATE 6.7

ourselves over. Scott ranged far and wide with his camera and his handshake, while I sat near the Silva family, stuffing my face with an onion-laden hot dog, making small talk, and trying to take their measure. Perhaps both my troubles could be solved at once.

"Say," I said, "I won that painting over there, the one with the turkeys…"

"That's a beautiful painting," Dan said. "I never win anything."

"It *is* a beautiful painting. But you see, I have this tiny apartment in the Bay Area and don't really have any place to put it. So, if you want it, you can have it."

Dan was genuinely moved by the gift.

"I have the perfect place for it," he said.

Thus, Scott and I were invited over to the Silva home in Elk Grove for a home-cooked meal of pheasant.

Megan Dipino and Savannah Shoup tore savagely at the breast feathers of two pheasants and giggled. Perhaps eight years old, these two girls had taken over the cleaning station, and their hands were covered in blood and feathers. An adult volunteer watched them bemusedly and kept them away from the knives.

"Here," he said, slicing open the breast of one of the birds. "This is the heart."

"Can I keep it?" asked Savannah Shoup.

"Sure," he replied, reaching for a plastic bag.

"My mom said I couldn't go hunting until I cleaned a pheasant. She didn't think I would like it. I want to show her," she said.

"Well, here's another thing you can do. See the white thing sticking out of the leg? That's a tendon. Watch what happens when I pull it," he said, making the pheasant's claw open and close. The girls were suitably excited by this development and collected some feet to join the heads, hearts, and tail

feathers in their bags. I grinned. That trick was familiar to me, though I hadn't thought of it for almost twenty years. In autumn a few boys would bring pheasant feet into school. The best trick was to use them to pull the hair of girls that stood in front of them in choir. I doubt a single one of those boys kept possession of a pheasant leg until the end of the day, and if one did, shame on him.

I was plagued by theories of pedagogy and determinism, and so I grabbed a couple pheasants and washed the guts

---

The crowd exuded pride, anticipation, caution, responsibility, and the unmistakable radiance of hunter's orange.

---

out of their chest cavities, tossing the unwanted innards into a nearby trash barrel, barely the master of my rebellious gorge. Why could these two girls so gleefully pluck and disembowel these birds while I shook with repressed weeping and tried not to vomit? I had always been thus. Why? And why, if I was built to resist this moment, did I take such sure and sensual pleasure in eating and preparing meat? All the vegetarians I have known claim they never really liked meat and that it was easy to give up. Everything smelled of death, of rot, and of wrong. I fled.

I washed my hands. They smelled like death. We said our good-byes and left the pheasant hunt. I stopped at a gas station and washed my hands. They still smelled like death.

"My hands smell like death," I told Scott.

"You mean like meat?"

"No. Like death."

"I don't smell anything. I think it's one of those 'Out! Out! Damned spot!' smells."

"Hm."

Plate 6.8

PLATE 6.9

The Silvas live in a comfortable neighborhood riddled with cul de sacs in Elk Grove, just south of Sacramento. Their yard and the yards of their neighbors sported signs that said SILVA FOR MAYOR.

"Dan," I said, as we sat down in his kitchen and watched his wife prepare dinner, "are you running for mayor?"

He was, and on an antidevelopment ticket. Good man!

They fed us wine and introduced us to Grandma. Dan's son had scoured the Internet for the pheasant recipe he liked best. His choice would have saddened but not surprised Scott Leysath. The recipe called for the pheasant breasts to be pounded flat, breaded, and then baked in about three pounds of honey and butter seasoned liberally with black pepper. In case the gamey pheasant didn't agree with someone, Dan's wife also prepared some chicken.

The warmth, comfort, company, and wine of the Silva home turned off my writer's mind and I let myself become just a guest for a while. The food was delicious, the pheasant was more popular than the chicken, and the conversation bubbled and sparkled. I held my wine glass by the bottom of the stem to keep my hands far from my nose.

Plate 6.10

Plate 6.11

Everyone Should Know Everything

OUR FORTIETH AND FINAL DAY IN THE VALLEY began in darkness. The night seemed plated onto the sky, punctured only slightly by the uneven headlights of the Jetta as we rounded the curves of the mountain roads towards North Fork. Our hearts were heavy with expectation. This day was the terminus of our journey, the last bite in this gluttonous feast.

And to find the Valley we had been compelled to leave it. We owed honor to Geography. It had supplied the warp through which we ran the threads of our weaving. But what honesty it was to be free of its poor conceit—that a land known was a land parsed. Rivers and roads, fires and floods, birds and beasts are promiscuous in their peregrinations, or, better yet, faithful to a higher idea. The mountains, the foothills, the valley, the forests, deserts, and seashores, while distinct, are not separate, and remain nameless without the variety of their brethren. There is an idiocy at the base of our sciences, but luck had drawn us away from its traps. In the foothills can be found an ancient people who once inhabited both peak and plain. Sandy Clark, of the North Fork Mono, would be waiting for us just after sunrise, and with her the leaching of acorns.

A small shard of weak dawn light fell onto a meadow to my left, illuminating the fiery leaves of a single autumnal tree, while all around the meadow crowded the dim needles of conifers. I said nothing, secretly believing in the profanity of speech, and let the burning vision slide past as we continued our ascent. I knew, however, that it had come to comfort me with the terror of its beauty. This journey, and now this book were ending, and we were right on time.

We drove past the North Fork Mono Museum and took a left after the old lumber mill. The road narrowed and fractured into dead ends and driveways as we cleaved the densely wooded mountainside, finally arriving at an impressive wooden home with a spacious observation of the undulating decline of the foothills beyond.

Sandy was ready for us and offered coffee. We sat in the kitchen surrounded by bookshelves filled with science fiction and fantasy. Sandy and her daughter rent a part of the house from another woman and the books were hers. In the backyard, the green ribbon of a garden hose led to a small wooden square elevated by cinder blocks. I don't remember what we talked about, but I recall the warmth of the coffee cup against my palms.

We had accepted an invitation to the Sierra Mono Museum the Friday before. Had we been able to turn up on Thursday, we would have been greeted by a roomful of elders cracking acorn shells and pounding the meat into meal, but on Friday, only Sandy Clark agreed to talk to us, in between her duties as a school bus driver. "Some Indians are very careful about who they give knowledge to, but I'm different. I think everyone should know everything."

Acorns are usually processed around a picnic table behind the museum, but we helped Sandy carry the nuts and tools up a gentle path to an ancient pounding rock. The rock had a shallow indentation off to one side, created from generations of use. The shelled acorns would be

the cracking surface to a comfortable height, and, as the acorn cavity had been shaped by the acorns themselves, it was a perfect fit. They did not speak like rocks, once you knew them. They spoke like tools.

The economy and providence of the stone tools were impressive and unexpected, but the soaproot brush was poetry to their prose. The brush itself was made from the bound black fibers that had once surrounded a single root, and the handle was formed by layer upon layer of a sticky paste made by pounding the root. Because the handle was molded by a human hand, it fit comfortably inside a human hand. The brush had two distinct sides; the ends all curved in one direction like a cresting wave. This side

---

Sandy Clark, of the North Fork Mono, would be waiting for us
just after sunrise, and with her the leaching of acorns.

---

pounded to meal within the indenture. The rest of the tools consisted of two more rocks and a soaproot brush. One rock was soft, with a tiny, acorn-sized hole near the center. Sandy placed the narrow end of an acorn into the hole and hefted a round, hard stone in her other hand. One quick tap on the top of the acorn split the shell into several pieces and left the meat beneath perfectly intact. Sandy cracked a handful of nuts this way and then passed the rocks over to me.

I essayed a hesitant tap on an acorn and was pleased to find the shell split open just as they had for Sandy. What I had seen from afar as rocks, no matter how cleverly they were being used, in my own hands became tools as perfectly conceived and designed as the most artful Japanese knife. The cracking rock fit snugly into my closed hand, not round, but rounded, and smooth. It fit so that many hours could go by without my hand tiring of its augmentation. The soft soapstone of the rock that held the acorn elevated

was perfect for gathering large pieces of acorn or making quick sweeps of a large area. The other side, created by the smooth back of the wave, allowed for more detailed work and let not even the smallest bits of acorn escape its ministering.

We were not cracking this year's acorns. Acorns are gathered in the fall and must age and dry for a year before they can be cracked and pounded. Traditionally, the Mono built granaries of woven branches and set them about a foot from the ground, but burlap sacks seem to work just as well. There are six native oak trees that provide acorns to the Mono people, but they don't bear nuts every year. The Mono must know the pattern of each tree to know what year they will harvest it. Providence has ensured that not a year goes by without a harvest.

After the demonstration, we continued walking with Sandy up a trail that leads to a small replica of a native village. Sandy named each plant and tree we traveled past,

PLATE 6.12

Plate 6.13

slowly humanizing the forest landscape. Near the edge of the replica village stood a tall pine whose thick bark was completely pocked with small holes.

"The woodpeckers drill those holes," Sandy explained, "and then pick up acorns we have dropped and stick them in there. After a while a worm gets inside each acorn. The woodpeckers let the worms grow fat, and then they break the acorns open and eat them." The tree contained hundreds of holes and must have constituted a veritable worm ranch for the enterprising woodpeckers.

To walk through ancestral lands with a person native to that place is to obliterate wilderness. No longer an expanse of untouched or unspoiled innocence, the natural landscape becomes a network of human industry. It is ranch, mine, farm, and toolshed all at once. What was mute to me had names for her. I could not help but wonder what our country would look like if we had built upon that treasury of knowledge and that way of knowing rather than seeking its obliteration and importing the trappings of our abandoned landscapes.

We followed the garden hose through clumps of buffalo grass. A number of half-grown cats chased each other ferociously across the lawn, up trees, and all over the back porch. To leach the acorn meal, Sandy had stretched a square of canvas over a wooden frame and set it on the aforementioned cinder blocks almost a foot above the ground. The meal was placed in the canvas. The garden hose dumped water into the meal along one edge, a fresh bough of cedar filtering the water and dispersing its flow. Underneath the canvas, the escaping water fell in a gentle torrent on the ground and into a metal can.

"It looks almost ready," Sandy said, surprised. "Usually it takes a few more hours than this, but then, I didn't put so much in."

She turned off the water and dipped a finger into the soft white meal, beckoning for us to follow her example.

They did not speak like rocks, once you knew them. They spoke like tools.

The taste is gentle and hard to describe. It is more glutenous than most flour and contains a mineral trace. It isn't a bit bitter, though.

"Uh-oh. See that dark spot there? We'd better try it again."

We each dipped our fingers near the dark spot and tasted again. Grimaces and spitting followed soon after.

"Looks like it needs more time," Sandy said, resuming the flow of the water.

Once the meal is ready, the cook places it in a special basket for cooking and adds water. A spoon made from a willow branch holds a round rock that has been sitting in the fire. The cook places the hot rock in the flour and stirs it around until it loses its heat. Another is added, and the process continues. Eventually the flour and water mixture is thoroughly cooked and, *voila*: Acorn Soup.

"The best part is the cracker that forms around the cooking rock," Sandy said. Our timing was such that we were not able to see the famous native soup being made, nor to sample it. But Sandy let us taste some manzanita berries and some sourberries before we left. The manzanita berries had a leathery flesh that was sweet and salty, while the sourberries lived well up to their name.

THE VISION WAS ALREADY ENDING when we pulled into Academy, a town on its last legs located just where the Valley begins to forget the Sierra Nevada foothills. While we waited for North Fork Mono and fallow-deer rancher Ron Goode to ring us up, we dozed in the shade around an old, locked-tight Methodist church breathing in and out a kind of fatalism. We grew silent together. The end. The end of this trip, this project, and this book. Acorns and deer. One should probably not live books, I thought, as I lay in the wood-between-worlds of a ghost town church.

Ron finally called but had little time to talk. He was due at a wedding in less than an hour, but he took us around his pens and showed us the soft hide of a deer stretched and curing in the sun, and we petted one of his pets. I asked him what a deer was like as a pet.

"A pain in the ass," he said.

Ron was ruminating on the paradoxes of his existence, that he has domesticated a non-native deer because too few native deer can be hunted to serve the needs of the native any. I had one buck that I kept penning up for three years before he was finally ready. Well, I better get going."

We were stunned as we left Ron, and I kept repeating what he had said, in order to take it in fully. If there was some process by which an animal could accept its death, and if we only used those animals, then this guilt, this spot I carried with me, might finally out. But we didn't see the thing, we were not witnesses but just the recipients of some words and this singular idea, that animals may give their death. It's better by far than what we die for, banners and flags and systems of government.

As we made the long drive home I tried to trade in the microscope of edibilism for something more telescopic.

I could not help but wonder what our country would look like if we had built upon that treasury of knowledge...

---

community, and that his ranching is a thorn in the side of a variety of government agencies, when his daughter, Elena Littlefawn, came out and fixed him with one of her many looks that could kill. Her beauty startled me into ignoring her completely. It was time for Ron to go. Past time. He continued talking a little longer as a token assertion of his paternal authority.

"How do you slaughter the deer?" I asked him, unconsciously using a domesticated term.

"One of the problems with the park deer," Ron said, "is that they no longer want to die. We will hunt them, but then when it is time to take them, they are frightened, and we cannot kill them. Here, I choose three or four deer and put them in a separate pen for a few days. I watch them very carefully, and after a while, all but one will calm down. The chosen deer will stomp its feet on the ground and then urinate and lie down. Once the deer is ready to die, we kill it. Sometimes none of the deer are ready and we don't kill

All of the people we had talked to were trying to get stuck wherever they were. The land demands this from us: to become native. Using the fullness of our engagement with food, from imagined orchards to following a deer's eye-like tracks to the actual handwork of preparing food, the smells and tastes and the bringing of others to us for it, could it be that we are actually trying to come to peace with the land—the land to which almost all of us are still strangers? Or, by following the more common practice of taking the power of food and denuding it in an industrial system that annihilates place, are we reducing the guilt we feel for the way we live?

The cost of our exile is great, but greater still is the knowledge that buried within so many of us is the heedless, speechless compulsion to pick berries from the roadside when we see them, and to remember that place, and to tell our children where they should go, and how they should go about it.

PLATE 6.14

Since its founding in 1974, Heyday Books has occupied a unique niche in the publishing world, specializing in books that foster an understanding of the history, literature, art, environment, social issues, and culture of California and the West. We are a 501(c)(3) nonprofit organization based in Berkeley, California, serving a wide range of people and audiences.

We are grateful for the generous funding we've received for our publications and programs during the past year from foundations and more than three hundred and fifty individual donors. Major supporters include: Anonymous; Audubon California; BayTree Fund; B.C.W. Trust III; S. D. Bechtel, Jr. Foundation; Fred & Jean Berensmeier; Joan Berman; Book Club of California; Butler Koshland Fund; California State Automobile Association; California State Coastal Conservancy; California State Library; Candelaria Fund; Columbia Foundation; Community Futures Collective; Compton Foundation, Inc.; Malcolm Cravens Foundation; Lawrence Crooks; Judith & Brad Croul; Laura Cunningham; David Elliott; Federated Indians of Graton Rancheria; Fleishhacker Foundation; Wallace Alexander Gerbode Foundation; Richard & Rhoda Goldman Fund; Marion E. Greene; Evelyn & Walter Haas, Jr. Fund; Walter & Elise Haas Fund; Charlene C. Harvey; Leanne Hinton; James Irvine Foundation; Matthew Kelleher; Marty & Pamela Krasney; Guy Lampard & Suzanne Badenhoop; LEF Foundation; Robert Levitt; Dolores Zohrab Liebmann Fund; Michael McCone; National Endowment for the Arts; National Park Service; Philanthropic Ventures Foundation; Alan Rosenus; Mrs. Paul Sampsell; Deborah Sanchez; San Francisco Foundation; William Saroyan Foundation; Melissa T. Scanlon; Seaver Institute; Contee Seely; Sandy Cold Shapero; Skirball Foundation; Stanford University; Orin Starn; Swinerton Family Fund; Thendara Foundation; Susan Swig Watkins; Tom White; Harold & Alma White Memorial Fund; and Dean Witter Foundation.

For more information about Heyday Institute, our publications and programs, please visit our website at www.heydaybooks.com.

Other BayTree Books

BayTree Books, a project of Heyday Institute, gives voice to a full range of California experience and personal stories.

Walking Tractor: And Other Country Tales (2008)
Bruce Patterson

Archy Lee: A California Fugitive Slave Case (2008)
Rudolph M. Lapp

Where Light Takes Its Color from the Sea: A California Notebook (2008)
James D. Houston

Tree Barking: A Memoir (2008)
Nesta Rovina

Ticket to Exile: A Memoir (2007)
Adam David Miller

Fast Cars and Frybread: Reports from the Rez (2007)
Gordon Johnson

The Oracles: My Filipino Grandparents in America (2006)
Pati Navalta Poblete

About the Author and Photographer

WILLIAM EMERY grew up on a small farm in the Smoky Hills of Kansas. He studied Russian literature and religious studies at the University of Kansas and is a former editor at Heyday Books. He splits his time between the San Francisco Bay Area and Kansas. He is currently working on a novel, a collection of prose poetry, and a children's book.

SCOTT SQUIRE is a documentary photographer working at the intersection of journalism, anthropology, and fine art. A fourth-generation son of the Pacific Northwest, Squire was educated at the University of Washington (anthropology) and the University of California, Berkeley (visual journalism). He is a principal at NonFiction Media, a social justice–oriented multimedia production company based in Seattle, Washington.